AES-Vol. 26

HEAT PUMP DESIGN, ANALYSIS, AND APPLICATION

presented at
THE WINTER ANNUAL MEETING OF
THE AMERICAN SOCIETY OF MECHANICAL ENGINEERS
ATLANTA, GEORGIA
DECEMBER 1–6, 1991

sponsored by
THE ADVANCED ENERGY SYSTEMS DIVISION, ASME

edited by

K. E. HEROLD
UNIVERSITY OF MARYLAND

V. MEI
OAK RIDGE NATIONAL LABORATORY

P. E. SCHEIHING
U.S. DEPARTMENT OF ENERGY

THE AMERICAN SOCIETY OF MECHANICAL ENGINEERS
United Engineering Center ■ ■ ■ 345 East 47th Street ■ ■ ■ New York, N.Y. 10017

ISBN No. 0-7918-0869-6

Library of Congress
Catalog Number 91-58434

FOREWORD

This volume consists of 12 papers which span a wide range of interests related to heat pump technology. The papers were submitted for the symposium on Heat Pump Design, Analysis and Application of the 1991 ASME Winter Annual Meeting. The symposium is sponsored by the Heat Pump Technical Committee of the Advanced Energy Systems Division of ASME.

The Heat Pump Technical Committee (HPTC) was founded in 1987 by Dr. Fang C. Chen who acted as the first Chairman. The HPTC is currently led by Dr. Keith E. Herold of the University of Maryland, Chair, Dr. Karen den Braven of the University of Idaho, Vice-Chair, Dr. Vince Mei of Oak Ridge National Laboratory, Paper Review Coordinator and Dr. Lin-Tao Lu of Rheem Manufacturing Co., Secretary. The HPTC has grown considerably during 1990–1991 and we anticipate continued growth because of renewed interest due to many factors. We invite participation from anyone interested in heat pump technology. We are particularly interested in attracting engineers from industry. Anyone interested in heat pump technology is welcome to attend our meetings (yearly meetings at ASME WAM) and participate in our activities (main activities are sponsoring technical sessions and bestowing best paper awards).

The Heat Pump Technical Committee exists to facilitate development of widely diverse heat pump technologies. These include vapor compression, absorption, adsorption, magnetic and many other technologies. Interest in alternative heat pump technologies has been growing in recent years with increased environmental concerns. The phase out of CFC's has caused a major disruption in the vapor compression heat pump industry. Concern about the greenhouse effect has encouraged renewed interest in energy efficiency. Both the CFC issue and the greenhouse effect concern have led to consideration of heat pump alternatives.

Keith E. Herold
Department of Mechanical Engineering
University of Maryland
College Park, Maryland

Vince Mei
Oak Ridge National Laboratory
Oak Ridge, Tennessee

Paul E. Scheihing
Office of Industrial Programs
U.S. Department of Energy
Washington, D.C.

LIST OF REVIEWERS

The following individuals donated their time and talents as reviewers for the papers in this symposium volume. The quality of the papers is a reflection of the efforts made by the reviewers and the session organizers would like to express our appreciation to each reviewer.

M. Ally
K. Amrane
V. D. Baxter
R. Chappell
S. Chaturvedi
J. Conklin
K. den Braven
L. A. Howe
D. Jung
L. Kirol
L.-T. Lu
M. McLinden
W. A. Miller
S. Morgan
R. W. Murphy
H. Perez-Blanco
S. Priebe
M. Rane
A. Rutz
S. V. Shelton
J. A. Shonders
W. H. Wilkinson
W. M. Worek

CONTENTS

GENERAL TOPICS IN HEAT PUMP DESIGN AND APPLICATION

Transient Dehumidification Characteristics of a Heat Pump in Cooling Mode
Srinivas Katipamula and Dennis L. O'Neal 1

Inexpensive Method for Performing Continuous Duty Cycle Testing of Engine-Driven Gas Heat Pumps
G. A Nowakowski, T. D. Haubert, and J. E. Lambert 11

A New Correlation of Adsorption Equilibrium
Lianquan Liu, Dhanes Charoensupaya, and Zalman Lavan 19

Effective Thermal Conductivity in Wet Desiccant Bed With Inert Materials
M. K. Sarwar and P. Majumdar 25

HEAT PUMP CYCLES AND WORKING MEDIA

Coupled Heat Transfer and Thermodynamic Adsorption Heat Pump Analysis
Daniel J. Miles and Sam V. Shelton 33

Design Optimization and Development of an Energy-Efficient Vapor Compression Cooling System
R. G. Mokadam, D. C. Nguyen, F. L. Sindermann, M. P. Desai, and D. B. Wigmore 39

Conceptual Studies on Modular Adsorption Systems for Various Applications
K. F. Knocke, A. Gerlich, and R. Strauss 49

Study of Thermoelectric Technology for Automobile Air Conditioning
B. Mathiprakasan, P. Heenan, V. C. Mei, and F. C. Chen 57

INDUSTRIAL PROCESS HEAT PUMPING AND REFRIGERATION

Practical Problems in the Implementation of Heat Pump in the Synthetic Rubber Industry
Deepak Shukla, Rajiv Singh, and Dennis Perkinson 63

High-Temperature Absorption Heat Pumps in Industry
Lawrence A. Howe and Donald C. Erickson 69

Heat Pump Applications in a Corn Processing Plant
Andrew S. McMullan 75

Measurements of Heat and Mass Transfer Coefficients During Absorption of Water Vapor by Lithium Bromide and (Li, K, Na)NO_3 Mixtures
A. Zaltash, M. R. Ally, R. L. Linkous, and L. N. Klatt 83

Author Index 91

TRANSIENT DEHUMIDIFICATION CHARACTERISTICS OF A HEAT PUMP IN COOLING MODE

Srinivas Katipamula and Dennis L. O'Neal
Department of Mechanical Engineering
Energy Systems Laboratory
Texas A&M University
College Station, Texas

ABSTRACT

Much of the seasonal cooling operation of the heat pump occurs at part-load conditions when the unit cycles on and off to meet the cooling load. The seasonal efficiency under part-load conditions of the heat pump is typically estimated from a laboratory measurement of the degradation coefficient (C_D). Manufacturers are only required to estimate C_D at a single test condition where the indoor coil performs sensible cooling only. The effects of transient dehumidification losses are not accounted in estimating the seasonal efficiency. In hot and humid climates, dehumidification performance of a heat pump is as important as the sensible cooling performance. Therefore, a series of tests were designed to quantify the part-load dehumidification characteristics of a 3-ton residential air-to-air heat pump.

The tests include: cycling rates from 0.8 to 10 cycles per hour (cph), percent on-times of 20, 50, and 80 %, indoor dry-bulb temperature between 22.2 °C and 26.7 °C, and indoor relative humidity between 20 to 67%. The outdoor conditions and the indoor air flow rate were constant for all test runs. All experiments were performed in psychrometric chambers under controlled conditions.

The dehumidification process started between 60 to 150 seconds after start-up depending on the test conditions. During start-up, the losses in the latent capacity were greater than the losses in the sensible capacity. The dehumidification response increased with indoor dry-bulb temperature at constant relative humidity and decreased with indoor temperature at constant dew-point temperature.

INTRODUCTION

In the cooling mode the heat pump absorbs heat from the air inside a residence and rejects it to the outside air. In the process it performs two functions: sensible cooling and dehumidification. In hot and humid climates the dehumidification capabilities of the heat pump are important in achieving and to maintaining comfort in the conditioned space [Katipamula et al., 1987, 1988].

Upon compressor start-up, the cooling capacity of a heat pump increases to steady-state over several minutes. The slow response leads to average capacities and efficiencies which are lower than the steady-state values. Since heat pumps operate at a part-load condition for long periods of time, understanding their transient dehumidification response is essential to quantify both comfort and efficiency.

There are number of variables which affect the transient performance of a heat pump: (i) percent on-time, (ii) the thermostat cycling rate, (iii) the indoor dry-bulb temperature, (iv) the outdoor dry-bulb temperature and (v) indoor relative humidity. In this study the affects of percent on-time, cycling, indoor dry-bulb and humidity on the transient dehumidification performance of a heat pump are quantified.

LITERATURE REVIEW

A number of researchers have recognized the significance of the transient losses in capacity and efficiency at start-up (Kelly and Bean 1977; Parken et al., 1977; Bullock and Reedy, 1978; Murphy and Goldschmidt, 1979; Hart and Goldschmidt, 1980; Baxter and Moyers 1985; Miller, 1985, O'Neal and Katipamula, 1991, Katipamula and O'Neal 1991).

Kelly and Bean (1977), tested a 5-ton heat pump installed in a residence in Washington D.C. In cooling mode, the degradation due to part-load operation on performance reported to as high as 18%. The degradation of the latent capacity due to part-load operation was not reported.

Murphy and Goldschmidt (1979), noted that the losses in performance due to transient effects is most likely attributed to refrigerant dynamics rather than the thermal mass of the heat exchanger coils. They concluded that the degradation coefficient is not unique but depends on the transient response under a fixed set of test conditions. Hart and Goldschmidt (1980), suggested that the transient response is not a simple function of the thermal mass of the heat exchanger coils and that the degradation of performance was different for the heating and cooling modes of a heat pump.

Numerous researchers have attempted to study the transient response of heat pumps in a controlled environment: Parken et al. (1977), Murphy and Goldschmidt (1984), Bullock and Reedy (1978), Miller (1985), Katipamula and O'Neal (11).

Parken et al. (1977) studied the transient performance of a 10.6 kW (3-ton) heat pump in a laboratory for both heating and cooling at different cycling rates and loads. At 20% on-time (i.e. the on-time to total cycle time ratio), the decrement of the part-load cooling capacity from the steady-state cooling capacity changed from 10% at a cycling rate of 0.8 cph to 35% at 4 cph. For an on-time of 80%, the decrement in cooling capacity ranged from 4% at 0.8 cph to 8% at 4 cph. The compressor power was relatively unaffected by cycling rate.

Parken et al. (1985) evaluated the cooling performance of three heat pumps in the field. The results indicated that the cooling load and part load factor (PLF) were independent of the outdoor temperature and dependent on the cycling rate and the percent on-time. The study also found that

$$\frac{\dot{Q}^l_{ss}}{\dot{Q}^s_{ss}} \neq \frac{\dot{Q}^l_{cyc}}{\dot{Q}^s_{cyc}} \qquad (1)$$

where, the quantities $\dot{Q}^l_{ss}$ and $\dot{Q}^s_{ss}$ denote latent and sensible steady-state capacities, respectively and $\dot{Q}^l_{cyc}$ and $\dot{Q}^s_{cyc}$ the latent and sensible cooling done over a complete cycle. This shows that the cyclic latent capacity and cyclic steady state capacities reach steady-state at different rates. However, the authors could not quantify the latent dehumidification performance in the field because they were only measuring the drop in the dry-bulb temperature across the indoor coil.

Previous study by the authors (Katipamula and O'Neal 1991) indicated that the PLF can be as low as 0.65 at low percent on-times and high cycling rates. It took between 6 to 15 minutes for the system to reach steady-state depending on the operating conditions. The study also suggested that the relationship between PLF and cooling load factor (CLF) is non-linear. Since the relationship was non-linear the C_D at 50 percent on-time was lower than at 80 or 20 percent on-times (O'Neal and Katipamula 1991; and Katipamula and O'Neal 1991).

A number of conclusions can be drawn from the literature reviewed:

(1) The losses due to transient effects can be as much as 20%.
(2) It takes 6–15 minutes to achieve steady-state after start-up.
(3) The transient response is affected by the number of on-off cycles and percent on-time during each cycle.
(4) The mass of the heat exchangers affects transient losses.
(5) The off-cycle migration of the refrigerant from the condenser to the evaporator causes significant losses in capacity.
(6) The relationship between CLF and PLF is non–linear.
(7) The compressor power was relatively unaffected due to part load operation.
(8) The transient performance is independent of the outdoor temperature.

Although a number of researchers have recognized the importance of transient losses in heat pump operation, mostly it has been in the heating mode. In cooling mode, the research was focused to quantifying the effects of heat exchanger mass, the off-cycle phenomena on the transient sensible capacity, and effects of percent on-time and cycling rate on the cooling performance. One of the major issue that has not been addressed so far is transient dehumidification performance. Thus there appears to be a need to better characterize the transient dehumidification performance of heat pumps.

EXPERIMENTAL FACILITY

To characterize the transient dehumidification response of heat pumps over a range of operating conditions, a series of experiments with a heat pump were devised. All the tests were performed in the psychrometric test facilities constructed and maintained according to the American Society of Heating and Air-Conditioning Engineers (ASHRAE) specifications (1983). The rooms can simulate both indoor and outdoor conditions required for testing the performance of the split system heat pumps.

A nominal 10.6 kW (3 ton) capacity split-system residential air-to-air heat pump was used for testing. The details of the indoor and outdoor heat exchangers are tabulated in Table 1. The test heat pump had a 10.6 kW (3 ton) capacity reciprocating compressor. The schematic of the indoor and the outdoor coils is shown in Figure 1.

Temperatures and pressures were recorded at many locations in the test loop (Figure 1). In addition, the refrigerant flow rate, the air flow rate, the temperature drop and the change in moisture content of the supply air across the indoor heat exchanger were recorded. Other points where temperature was measured on the test loop include: surface temperature of the accumulator and compressor and surface temperature of the liquid and vapor lines going to each of the four circuits of the outdoor heat exchanger.

Table 1 - Details of Outdoor and Indoor Heat Exchange Coils.

	Outdoor Coil	Indoor Coil
Shape	Horse Shoe Vertical	Vertical
Number of Rows	Two Rows, Four Circuit	Four Rows, Four Circuit
Ref. Tubing	0.95 cm dia Copper	0.95 cm dia Copper
Fin Spacing	7.9/cm	4.7/cm
Fin Type	Wavy	Wavy
Passes	Thirty	Seventeen
Frontal Area	1.63 m^2	0.35 m^2
Rated Flow	53.8 m^3/min	35.4 m^3/min

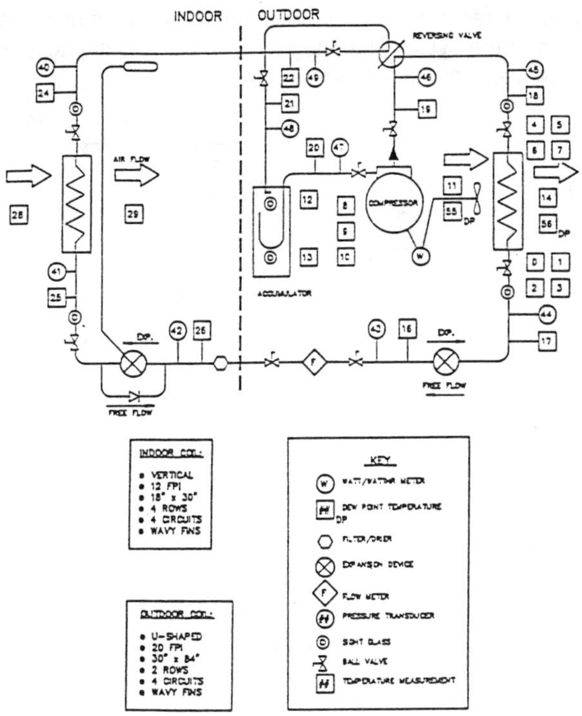

Figure 1 – Schematic of the Test Setup.

The sensor signals from the test points shown in Figure 1 were collected and converted to engineering units by a data logger. During each transient test of the heat pump, data processed by the data logger was transferred to a personal computer where it was stored. The fastest collection and storage rate for the cyclic testing was seven seconds per scan. The scan rate was adjusted such that at start-up the scan rate was seven seconds and gradually increased to 30 seconds after five minutes.

METHODOLOGY

The first step of the test procedure was to set the system refrigerant charge according to criteria specified by the manufacturer of the heat pump. The charge was set while the heat pump was operating in the cooling mode with an indoor dry-bulb temperature and humidity of 26.7 °C and 50%, respectively, and an outdoor dry-bulb temperature of 35 °C. The refrigerant charge was adjusted to obtain a 6.7 °C sub-cooling of refrigerant leaving the outdoor coil. The steady-state and cyclic tests recommended by the Department of Energy (DOE) were then run.

For all measurements, the psychrometric rooms were operated for more than one hour to allow the rooms and the equipment inside them to stabilize to the pre-set conditions before initiating a test run. Rooms were always maintained within the tolerance prescribed by the ASHRAE Standard (1983). For cyclic tests, three complete cycles were run to allow for consecutively repeatable cycles. As prescribed in the test procedure, all the cyclic tests except for the last cycle of a test were run with the continuous indoor fan operation (i.e. during both the on and off-cycles). The cooling capacity was measured by two methods: indoor air enthalpy and the refrigerant enthalpy. All measurements and calculations were in compliance with the ASHRAE Standard (1983).

The overall performance of the heat pump operating in cycles was characterized by measuring total cyclic capacity, total cyclic energy consumption, time required to reach steady-state, and cyclic energy efficiency ratio (EER_{cyc}). In addition to overall system performance, the details of the transient refrigerant and the air-side dynamics were analyzed carefully to completely understand the transient behavior of the heat pump.

EXPERIMENTAL TESTS

There are number of variables which affect the cyclic performance of heat pumps. This study included three different percent on-times (20, 50 and 80%) and broad range of cycling rates, including those which may be typically expected in a residence. It is often assumed that a typical residential thermostat cycles 1.8 cycles per hour (cph) at 20% on-time and 80% on-time and 3 cph at 50% on-time (1972). The indoor dry-bulb temperatures varied between 22.2 C and 26.7 C and indoor relative humidity varied between 20 to 67%. The outdoor conditions and the indoor air flow rate (34 m^3/min) were constant for all test runs.

TEST RESULTS

Steady-State

To compare the steady-state results with those from the manufacturer a base case steady-state test (Test A) was run. The indoor and outdoor rooms were maintained at 26.7 and 35 °C, respectively. The indoor relative humidity was maintained at 50% and the air flow rate was 34 m^3/min. The changes in the temperatures at steady-state condition were within the tolerance prescribed by the ASHRAE Standard (1983).

The total instantaneous cooling capacity obtained from the air enthalpy and refrigerant enthalpy calculation varied less than five percent during cooling operation. This agreement was within the ASHRAE test standard requirements (1983) and was observed in all test data.

Typical Cyclic Test Run

At start-up, the compressor reached steady-state speed almost instantaneously. Therefore, the suction and discharge pressures approached the steady-state values quickly (Figure 2). The refrigerant in the evaporator at start-up was saturated. The sudden drop in the suction pressure caused two phenomena. First, the refrigerant began to vaporize within the evaporator which rapidly cooled the evaporator. This phenomenon was characterized by the rapid increase in sensible capacity during the first 75 to 90 seconds after start-up (Figure 3). Figure 3 also shows the rate of moisture removal across the evaporator coil. Since two cycles were run before data were collected, the evaporator coil had moisture left at the end of the second cycle and was also at a lower temperature than the indoor room at the time of start-up. Therefore, the moisture removal is negative at start-up due to moisture being evaporated from the wet evaporator coil into the air stream.

The second phenomenon was the buildup of liquid refrigerant in the accumulator during start-up. This can be seen by studying the surface temperatures of the accumulator and the compressor (Figure 4). The compressor shell temperature, at the top, indicates that refrigerant inside the shell boiled due to sudden drop in suction pressure at start-up. Any refrigerant that boiled inside the compressor shell and any liquid that reached the accumulator was lost temporarily for cooling purpose, and contributed to the cycling losses.

After an initial rise, the power dropped to about 3.3 kW and remained constant (Figure 4). The initial rise in the power consumption corresponded to the boiling of refrigerant in compressor shell. The integrated cyclic cooling capacity and energy consumption were obtained by a numerical integration of the instantaneous values over the time of the cycle. The integrated power and capacity were then used to estimate the EER_{cyc}.

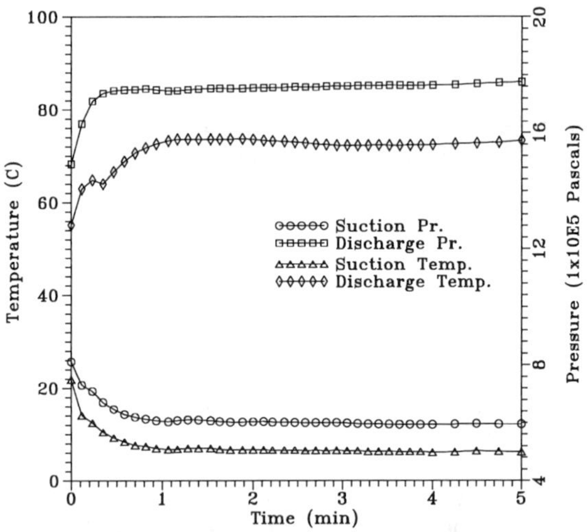

Figure 2 – Compressor Suction and Discharge Temperatures and Pressures.

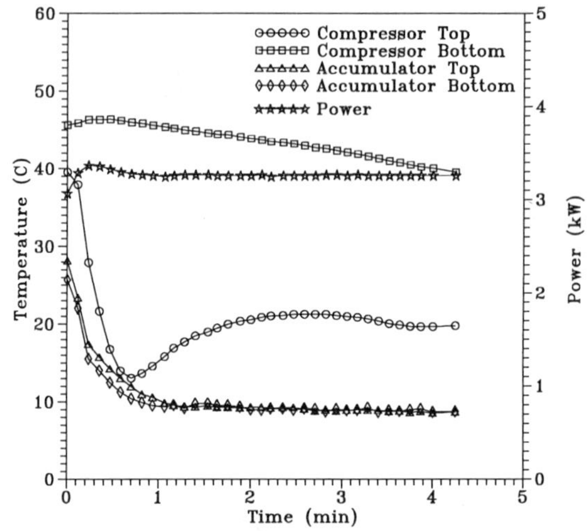

Figure 4 – Compressor and Accumulator Surface Temperatures.

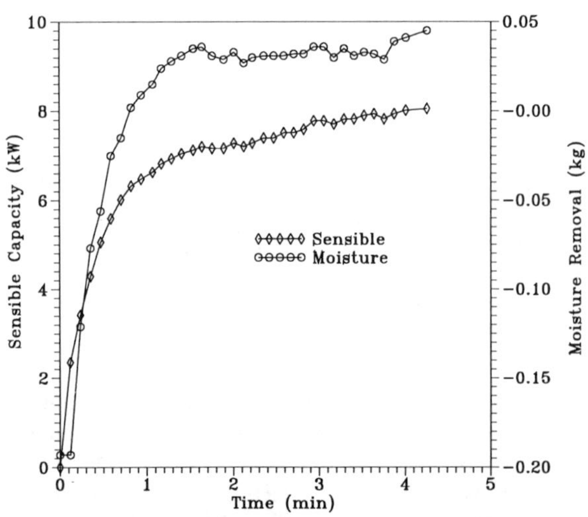

Figure 3 – Sensible, Latent, and Total Capacity for Typical Cyclic Tests.

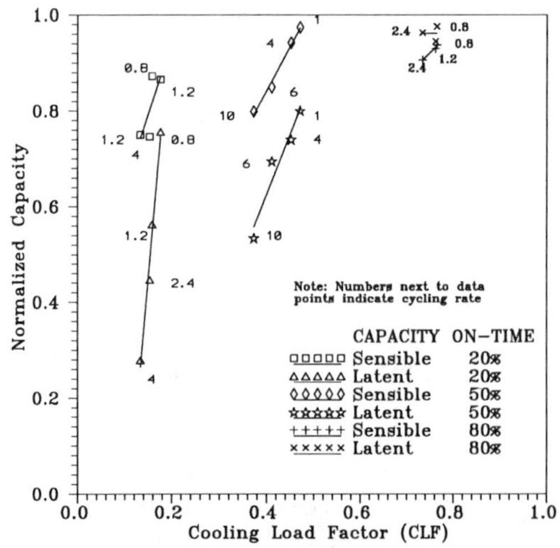

Figure 5 – Change of Normalized Sensible and Latent Capacities at Various Cycling Rates and Percent on-times.

RESULTS

One way to present results is in a non-dimensional or normalized form. For cycling effects on heat pumps, it is useful to normalize the latent (Q^L_{norm}) and sensible (Q^S_{norm}) capacities in terms of the steady-state values:

$$Q^L_{norm} = \frac{\frac{1}{t}\int_0^{t_{on}} \dot{Q}_L dt}{\dot{Q}_{Lss}}$$
$$Q^S_{norm} = \frac{\frac{1}{t}\int_0^{t_{on}} \dot{Q}_S dt}{\dot{Q}_{Sss}} \tag{2}$$

Figure 5 shows the normalized latent and sensible capacities at various percent on-times and cycling rates. The solid line represent the best fit through the data points. The normalized capacity is the ratio of the instantaneous capacity and the steady-state capacity at the same indoor and outdoor conditions as that of the cyclic test (Eq. 2). The loss of latent capacity due to cycling was greater than the loss of sensible capacity. As the cycling rate increased, the deviation of the latent capacity from the expected steady-state capacity increased. The deviation also increased as the percent on-time decreased. Although the effect of cycling rate and percent on-time on the sensible capacity was not severe as latent capacity the trends were similar. For instance, at 20 percent on-time and cycling rate of 4 cph the system lost over 70% for the steady-state latent capacity while the drop in sensible capacity was only about 15%.

The response of the moisture removal across the evaporator coil is shown in Figure 6 for three different indoor relative humidities (and 80% percent on-time). Initially, at start-up, the moisture removal was negative. Before condensation can occur on the evaporator, the surface temperature on the heat exchanger must drop below the dew-point of the air. When the system starts, the refrigerant temperature and the surface temperature of the evaporator are greater than the air dew-point temperature. Because there is moisture on the heat exchanger left from the previous cycle, some of this moisture will evaporate from the heat exchanger until the heat pump is able to reduce the evaporator surface temperature below the air dew-point. Thus, there is the negative moisture removal during the first 75 to 90 seconds (Figures 6 and 7). The steady-state moisture removal increased with increasing indoor relative humidity. The moisture removal reached steady-state quicker with increasing indoor relative humidity. It took 6 minutes, after start-up, for the moisture removal to reach steady-state for an indoor relative humidity of 67%, whereas it took almost 12 minutes at 45% Rh. There was no dehumidification below 45% Rh.

The moisture removal at 20% on-time is shown if Figure 7. Although the trends are similar to the 80 percent on-time case, the moisture removal becomes positive much earlier. It took less than 60 seconds for the moisture removal to become positive. Since the run time was only 7 minutes, the system never reached steady-state

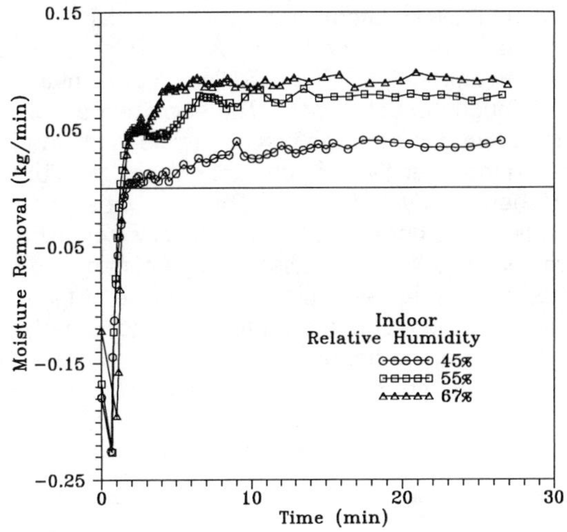

Figure 6 – Response of Moisture Removal Across the Evaporator Coil at Various Indoor Relative Humidities (80% on-time).

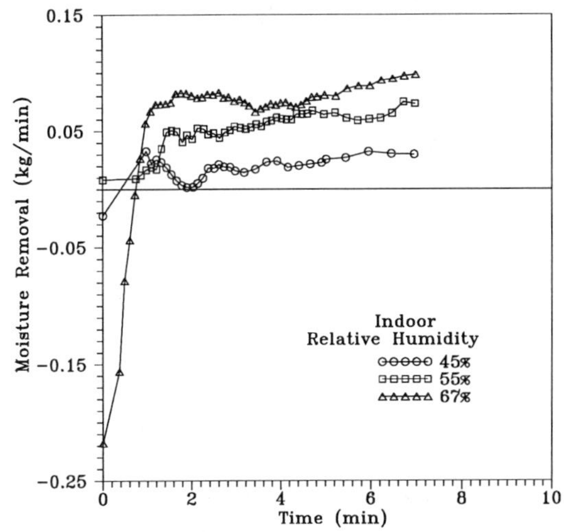

Figure 7 – Response of Moisture Removal Across the Evaporator Coil at Various Indoor Relative Humidities (20% on-time).

and the moisture accumulation on the evaporator coil was less than when the unit was run for 27 minutes (at 80% on-time). Therefore, the dehumidification process started early.

Dehumidification begins when the average surface temperature of the indoor coil falls below the dew-point temperature of the supply air. Since the dew point temperature of the supply air increases with increase in supply air relative humidity (at constant supply dry-bulb temperature), the potential for the dehumidification process also increases. Therefore, with increase in indoor relative humidity the dehumidification process will start much quicker. This leads to higher cyclic latent capacities. Although the moisture removal is negative for longer time period with increase in percent on-time, the cyclic latent capacity also increases with percent on-time.

The normalized latent capacity increased linearly with an increase in the indoor relative humidity (Figure 8) for a given percent on-time. The solid lines represent the best fit through the data points. The slope of the lines increased with decrease in percent on-time. With the increase in indoor relative (at constant indoor dry-bulb) the unit reached steady-state faster. Therefore, at a given percent on-time, the normalized latent capacity increased with relative humidity. With increase in the percent on-time, at constant indoor relative humidity, the run time of the unit increases and the unit has more time to reach steady-state. Therefore, the normalized capacities increase. The trends of the normalized sensible capacity were similar to that of the latent capacity (Figure 9). However, the slopes of the curve at a given percent on-time were smaller.

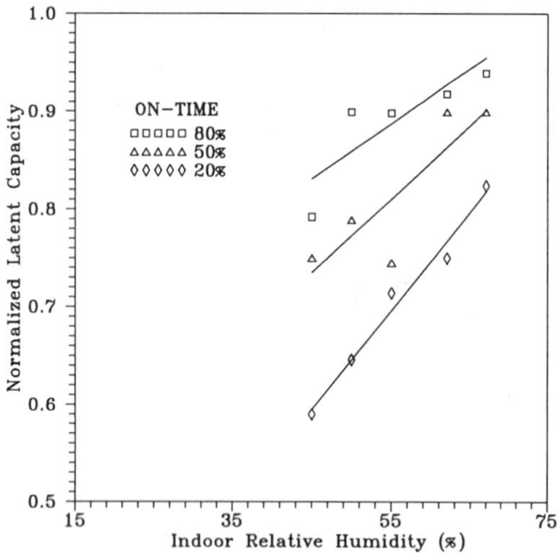

Figure 8 – Change in Normalized Latent Capacity With Indoor Relative Humidity.

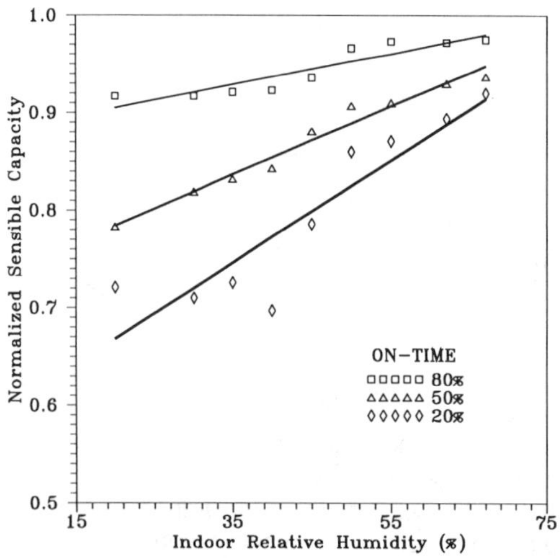

Figure 9 – Change in Normalized Sensible Capacity With Indoor Relative Humidity.

The response of the moisture removal at 50 percent on-time for various indoor dry-bulb temperatures and constant indoor relative humidity is shown in Figure 10. Moisture removal increased with an increase in the indoor dry-bulb temperature. Because moisture in the air-stream increases with an increase in indoor temperature at constant relative humidity, the moisture removal capacity also increases. The moisture removal was negative for the first 75 to 90 seconds after start-up. Similar to 50 percent on-time, the moisture removal increased with an increase in indoor temperature at 80 percent on-time. However, the moisture removal was negative for longer period of time (90 - 120 seconds).

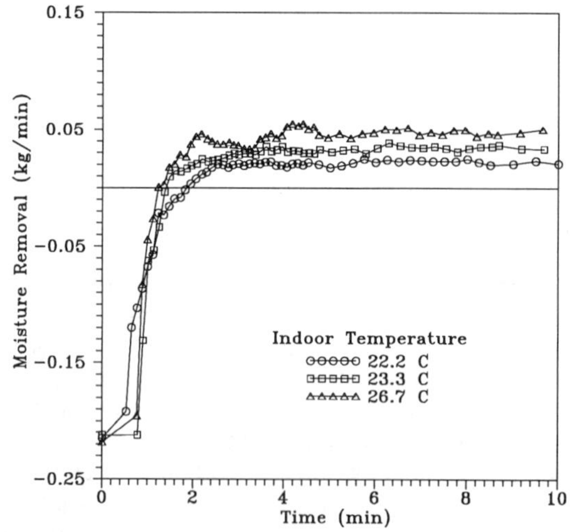

Figure 10 – Response of Moisture Removal at Various Indoor Dry-Bulb Temperatures and Constant Relative Humidity (50% on-Time).

The normalized latent capacity decreased with decreasing indoor temperature (Figure 11). In contrast, the normalized sensible capacity remained almost unchanged with indoor dry-bulb temperature for a given percent on-time (Figure 12). The normalized latent capacity showed larger slope than the normalized sensible capacity. The results for the 50 percent on-time showed a larger slope and variation than the 80% on-time. Since the run time of the unit at 50 percent on-time is smaller than at 80 percent on-time the slopes are larger.

At constant dew-point conditions, the moisture content in the air-stream is also constant. Therefore, to study the transient response of the heat pump under such conditions, a series of tests was run with constant indoor dew-point temperature (14.4 °C). The sensible capacity at the end of the cycle for both 80 and 50 percent on-times increased with an increase in indoor temperature. Moisture removal at the end of the cycle for both 80 and 50 percent on-times decreased with an increase in the indoor temperature. The time at which dehumidification started increased with an increase in indoor temperature. At 22.2 °C indoor temperature, it started after 60 seconds, whereas at 26.7 °C, it started after 120 seconds.

There was a slight increase in normalized latent capacity with indoor temperature for both 80 and 50 percent on-times (Figure 13). The normalized sensible capacity decreased slightly at 80 percent on-time and increased

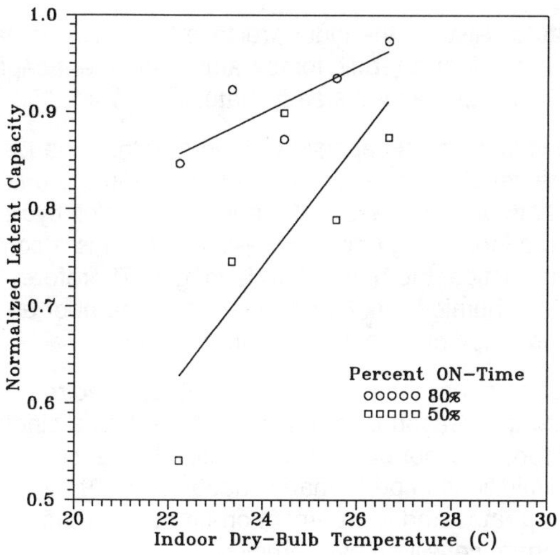

Figure 11 – Change in Normalized Latent Capacity With Indoor Dry-Bulb Temperature and Constant Indoor Relative Humidity.

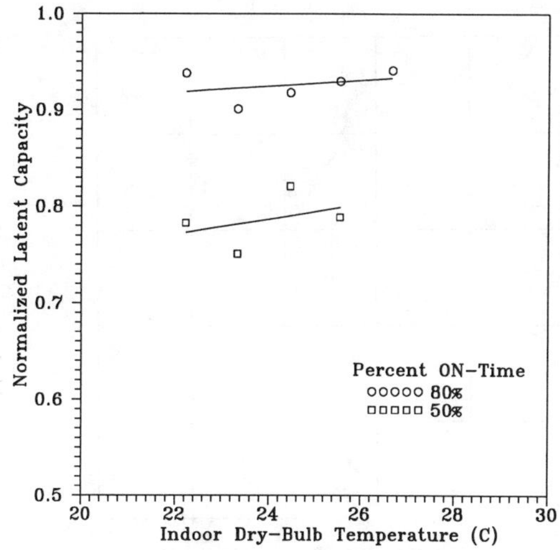

Figure 13 – Change in Normalized Latent Capacity With Indoor Dry-Bulb Temperature and Constant Indoor Dew-Point Temperature.

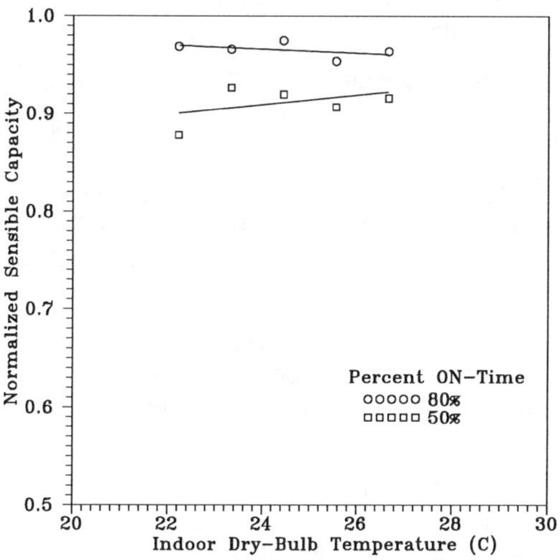

Figure 12 – Change in Normalized Sensible Capacity With Indoor Dry-Bulb Temperature and Constant Indoor Relative Humidity.

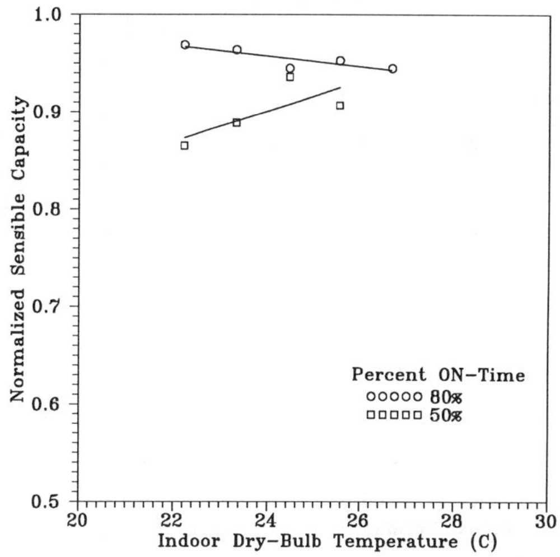

Figure 14 – Change in Normalized Sensible Capacity With Indoor Dry-Bulb Temperature and Constant Indoor Dew-Point Temperature.

slightly with indoor temperature at 50 percent on-time (Figure 14).

Finally, a relationship was developed between normalized capacity and percent on-time and indoor relative humidity. The capacity varies exponentially with percent on-time (Figure 3) and linearly with indoor relative humidity (Figures 8 and 9). Therefore, a non-linear regression analysis was performed on the experimental data. The regression model relating the normalized capacity to the percent on-time and indoor relative humidity is of the form:

$$cap_{nor} = 1. - exp(-\alpha \times pon) + \beta \times rh_i \times exp(-\alpha \times pon) \quad (3)$$

where, α and β are regression constants and pon and rh_i are percent on-time and indoor relative humidity, respectively. This model is consistent with previous models proposed by Goldschmidt and Murphy [6] and O'Neal and Katipamula [10]. This equation is valid for percent on-times greater than zero. The values for α and β which resulted in the best fit for the normalized latent and sensible capacity data are shown in Table 2. The value of α and β are system constants and may vary with the system. The predicted normalized latent capacity was in good agreement with the actual (Figures 15). The predicted normalized sensible capacity was also in good agreement; however, there was more scatter below 0.7 (Figure 16).

Table 2 - Regression Constants.

Capacity	α	β
Latent	1.391	1.121
Sensible	2.779	1.343

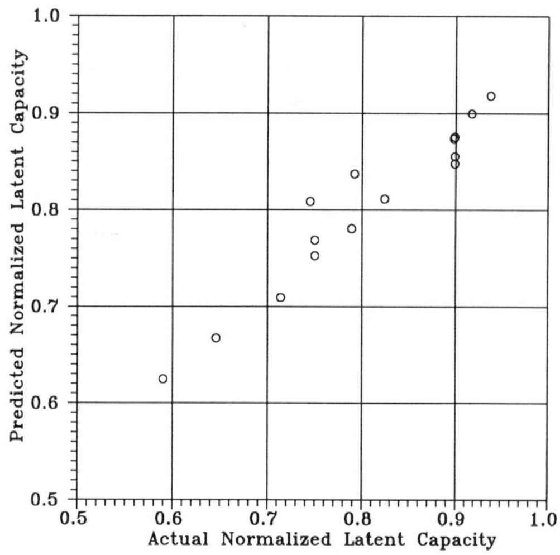

Figure 15 – Comparison of Predicted with Actual Normalized Latent Capacity

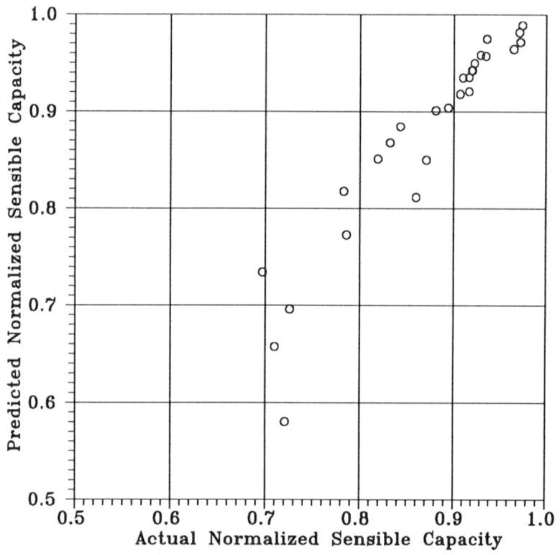

Figure 16 – Comparison of Predicted with Actual Normalized Sensible Capacity

CONCLUSIONS

To characterize the transient dehumidification performance of heat pumps in the cooling mode, an experimental and analytical investigation was performed. A series of tests were performed at varying operating conditions and the transient dehumidification performance was characterized. This study illuminated several trends required to characterize both transient sensible and dehumidification performances of the test heat pump:

(1) For almost all the tests, moisture was added at start-up and dehumidification began 60 to 150 seconds after start-up depending on the ambient conditions. Therefore, it appears that if the heat pump is ON for less 2 minutes it will add rather than remove moisture from the air.

(2) The latent capacity took 6 to 15 minutes to reach steady-state after start-up. As the on/off cycles increased it took longer for the latent capacity to reach steady-state. Also as the indoor relative humidity decreased (at constant indoor dry-bulb temperature) the latent capacity took longer to reached steady-state.

(3) The loss in latent capacity was greater than the loss in the sensible capacity with a decrease in the unit's run time and increase in the number of on/off cycles. If the conditioning unit is over-sized, which is a common practice, the unit will cycle longer. Therefore, the unit's dehumidification performance will be poor and the unit may not meet the comfort criteria.

(4) It appears that the current methodology used to estimate the seasonal performance, based on a single test run, may not be able to quantify the seasonal dehumidification performance adequately. At high cycling rates and low percent on-times the seasonal performance will be over predicted.

REFERENCES

AMCA/ANSI Standard 210–85, "Laboratory Methods of Testing Fans for Rating," 30 West University Dr., Arlington Heights, IL, 1985.

ASHRAE/ANSI Standard 116–1983, "Methods of Testing Seasonal Efficiency of Unitary Air–Conditioners and Heat Pumps," Atlanta, 1983.

Baxter, V.D., and Moyers, J.C., "Field–Measured Cycling, Frosting and Defrosting Losses for a High–Efficiency Air–Source Heat Pump," ASHRAE Transactions, vol. 91, pt. 2B, 1985, pp. 537–554.

Bullock, C.E., and Reedy, W.R., "Heat Pump Cyclic Performance and its Influence on Seasonal Operation," Third Annual Conference on Heat Pump Technology, Oklahoma State University, 1978.

Hart, G.H., and Goldschmidt, V.W., "Field Measurement of a Mobile Home Unitary Heat Pump," ASHRAE Transactions, vol. 86, pt. 2, 1980, pp. 347–367.

Katipamula, S., O'Neal, D.L., and Somasundaram, S., "Determination of the Transient Dehumidification Characteristics of High Efficiency Central Air–Conditioners," Report # ESL/87–04, Dept. of Mechanical Engineering, Texas A&M University, 1987.

Katipamula, S., O'Neal, D.L., and Somasundaram, S., "Simulation of Dehumidification Characteristics of Residential Air–Conditioners," ASHRAE Transactions, vol. 94, pt. 2, 1988, pp. 829 - 849.

Katipamula S. and O'Neal D.L., "Performance Degradation During on-off Cycling of Single Speed Heat Pump Operating in Cooling Mode: Experimental Results," ASHRAE Transactions, vol. 83 pt. 20.

Kelly, G.E., and Bean, J., "Dynamic Performance of a Residential Air-to-Air Heat Pump," National Bureau of Standards Building Science Series, BSS93, Washington, D.C., March 1977.

Miller, W.A., “The Laboratory Evaluation of the Heating Mode Part–Load Operation of an Air–to–Air Heat pump,” ASHRAE Transactions, vol. 91, pt. 2B, 1985, pp. 524–536.

Murphy, W.E., and Goldschmidt, V.W.,“The Degradation Coefficient of a Field Tested Self–Contained 3–Ton Air–Conditioner,” ASHRAE Transactions, vol. 85, pt. 2, 1979, pp. 396–405.

Murphy, W.E., and Goldschmidt, V.W.,“Transient Response of Air Conditioners — A Qualitative Interpretation Through a Sample Case,” ASHRAE Transactions, vol. 90, pt. 1B, 1984, pp. 997–1008.

NEMA 1972, "NEMA Standard for Low-Voltage Room Thermostats," Pub. No. DC3-1972.

O’Neal D.L. and Katipamula S., “Performance Degradation During on-off Cycling of Single Speed Air-Conditioners and Heat Pumps: Model Development and Analysis,” ASHRAE Transactions, vol. 83 pt. 20.

Parken, W.H., Beausoliel, R.W., and Kelly, G.E., “Factors Affecting the Performance of a Residential Air–to–Air Heat Pump,” ASHRAE Transactions, vol. 83, pt. 1, 1977, pp. 839–849.

Parken, W.H., Didion D.A., Wojuechowski, P.H., and Chein, L., “Field Performance of Three Residential Heat Pumps in the Cooling Mode,” NBSIR 85-3107, NBS, Washington D.C., 1985.

INEXPENSIVE METHOD FOR PERFORMING CONTINUOUS DUTY CYCLE TESTING OF ENGINE-DRIVEN GAS HEAT PUMPS

G. A. Nowakowski
Gas Research Institute
Chicago, Illinois

T. D. Haubert and J. E. Lambert
Battelle Memorial Institute
Columbus, Ohio

ABSTRACT

The use of recirculating air loops as a simple, cost-effective approach for durability testing of gas heat pumps is described. Several alternative test methods are also described and evaluated. These testing approaches provide a means for running the heat pump all available hours while simulating the actual running conditions expected in field service. The recirculating air loops have been operating employing two different duty cycles based on historical weather data for Chicago and Salt Lake City. The operation and limitations of the testing approach are also described.

INEXPENSIVE METHOD FOR PERFORMING CONTINUOUS DUTY CYCLE TESTING OF ENGINE-DRIVEN GAS HEAT PUMPS

INTRODUCTION

The Gas Research Institute is currently sponsoring work to develop a 5-ton engine-driven gas heat pump (GHP) for light-commercial applications. Aside from first cost and performance, two of the most critical factors affecting commercial success of space conditioning products are long life and high reliability. However, evaluation of system life and reliability requires extensive operating experience under expected "duty cycle" conditions.

Due to the excessive time required to accumulate actual operating experience in the field, it was necessary to develop a means of obtaining "accelerated" experience in order to gain confidence in system durability, reliability, and overall operation over a wide variety of operating conditions. A typical engine-driven GHP, on average, is expected to operate between 2,500 and 4,000 hours annually, whereas continuous year-round testing has the potential to subject the GHP to 8,760 hours per year or greater than two to three times that possible in the field. One approach, the recirculating air loop method, has been successfully implemented to gain operating experience with the engine-driven GHP over the shortest time-frame. The recirculating air loop method was chosen over other techniques that control engine and compressor loading due to considerations of cost and ease of testing. The primary benefit of this test method is the ability to simulate GHP operating conditions, thus enabling year-round continuous GHP operation under normal duty cycle conditions for most climates.

ALTERNATIVE HEAT PUMP LOADING SCHEMES

A number of approaches were considered in the selection of an endurance test method capable of subjecting the critical components of a GHP--the engine and compressor--to operating conditions expected in the field while accumulating operational time in the shortest time-frame. The most important factors in selecting the endurance test approach were:

- Accuracy of load control with minimal operator intervention and maintenance. The need for accurate load control can be seen by examining Figure 1, which shows a greater than two to one ratio of engine and compressor load over the range of outdoor temperatures. The lives of engine and compressor components are quite sensitive to the applied loads. For instance, life is inversely related to the cube of the load for rolling element bearings.

- Confidence in maintaining proper component operating conditions. For example, in the case of the compressor, it is important to avoid creating conditions such as excessive loading, liquid slugging or lubricant starving of the compressor. Testing of a complete heat pump helps to avoid such problems.

- Need to minimize development time and cost. Obtaining cost-effective operating experience in the shortest period of time is always a key factor in selecting an endurance test approach.

The simplest endurance testing approach is to install the GHP in a field application, heating and cooling a residence or building space. However, since operation of the GHP is dictated by weather conditions, operating hours are limited. Alternatively, the GHP

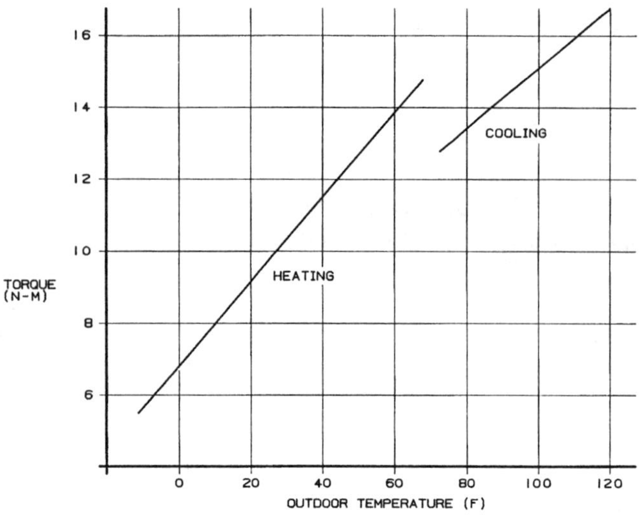

FIGURE 1. COMPRESSOR AND ENGINE LOAD-VERSUS-OUTDOOR TEMPERATURE FOR 5 TON GHP.

could be run continuously in a similar application. For instance, the indoor unit could be placed in a large building where its cooling or heating output does not significantly affect the building temperature. However, the additional operating hours would be predominantly accumulated at milder outdoor temperatures. Thus, the loading would not be representative of actual field conditions. Due to these considerations, it was decided that a means of simulating heat pump loading was needed.

Testing of the engine on a dynamometer and the compressor on a calorimeter is a good means of controlling loads. However, this approach was rejected because it was desired to test, at minimum, the engine and compressor, along with the belt coupling, as a complete system. Testing the indoor and outdoor units in environmental chambers is an excellent method for controlling operating conditions, but would be cost prohibitive for continuous long-term testing. Instead, three approaches that provide nearly the control of environmental chambers at much lower cost were considered. These were liquid baths, recirculating air loops and the hot gas bypass loop, which are described in the following sections.

Liquid Baths

The use of liquid baths to establish the condenser and evaporator temperatures for load control is illustrated in Figure 2A. The primary advantage of liquid baths is the ease of maintaining precise control of the bath temperatures. Of course, a non-freezing liquid, such as a water/glycol solution, would be required for the evaporator bath.

To maintain the evaporator bath temperature, a simple heating element could be used. However, for temperature control of the condenser bath, a large capacity chiller would be required. The chiller capacity must match the maximum heat rejection of the condenser, which is likely to occur at peak temperature cooling conditions. The chiller cost will generally be significant.

The standard heat exchanger coils for the heat pump may not be appropriate for liquid baths, since the fins will limit circulation of the liquid. If another heat exchanger coil is substituted, it is important to avoid large changes to the refrigerant flow passage, since this can affect the circulation and return of compressor lubricant. A safe approach is to use a coil similar to the heat pump coil in terms of tubing configuration, but without restrictive fins.

A standard thermostatic expansion valve could be used to maintain desired superheat conditions at the compressor inlet. There is no need for a reversing valve and defrost cycles are eliminated. While this simplifies the test system, it does mean that some heat pump functions will not be exercised.

One problem associated with this approach arises from the difference in heat exchanger efficiency in a liquid bath. In ordinary heat pump operation, there will be a significant temperature differential between the refrigerant and air. The differential will depend on coil design and both refrigerant and air flow rates. With a liquid bath the differential will be reduced, having an impact on compressor loading. At mild temperatures, the difference would be significant. If the heat pump characteristics, in particular the coil temperatures at various operating conditions, are well known, this effect can be compensated by adjusting the bath temperatures to achieve desired saturation temperatures in the coils.

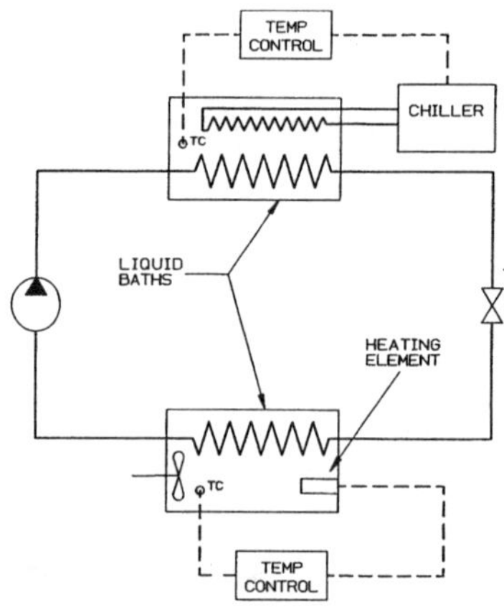

FIGURE 2A. COMPRESSOR LOADING WITH LIQUID BATHS

Another advantage of the liquid bath approach is the ease of measuring system performance. By insulating the baths and measuring the heat added to or removed from the bath, the system capacity and COP could be accurately monitored during normal operation.

Figure 2B shows an alternative approach that utilizes a heat exchanger between the liquid baths. With this approach the capacity of both the chiller and heating element can be reduced. However, it would increase the system complexity and make performance measurements more difficult.

A summary of advantages and disadvantages of this approach are:

Advantages

- Temperature control is accurate and reliable
- Performance measurement is relatively simple

Disadvantages

- Different heat exchangers may be required
- Cost and energy requirements for the chiller are significant
- Defrost cycles are eliminated
- Only engine and compressor are tested.

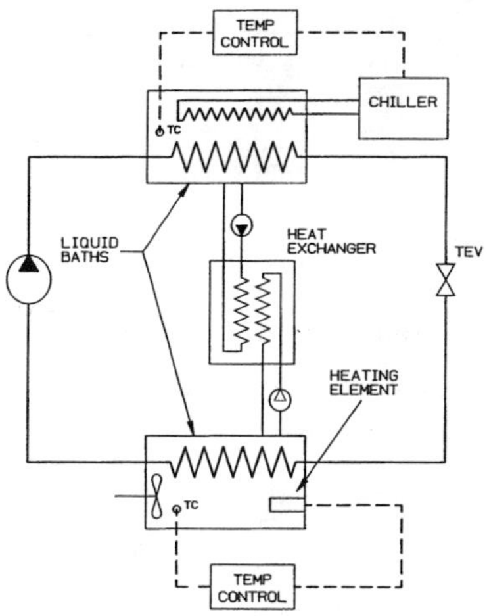

FIGURE 2B. LIQUID BATHS WITH HEAT EXCHANGER

Recirculating Air Loops

The use of recirculating air loops, illustrated in Figure 3, is applicable when complete heat pumps are available for testing. To be effectively operated year-round, both the indoor and outdoor units should be located in a heated building, though not necessarily air conditioned.

With this approach the heat pump loading is managed by controlling the inlet air temperatures to the evaporator and condenser coils. To illustrate its operation, assume we want to simulate heating operation at 30 F outdoor temperature and at 72 F indoor temperature, with the test room air temperature at 70 F. In the heating mode the outdoor unit is the evaporator and removes heat from the air. Initially, air within the outdoor unit enclosure is continuously recirculated over the evaporator coil, gradually decreasing the air temperature. As the evaporator inlet air temperature approaches 30 F, fresh test room air (at 70 F) is mixed with the recirculated air. The relative proportions of fresh air and recirculated air are adjusted until equilibrium is reached at the desired temperature. A similar procedure occurs at the indoor unit, although the air temperature needs to be increased only slightly above room temperature. In this case, only a small portion of the discharge air is recirculated.

In the cooling mode the outdoor unit acts as the condenser rejecting heat to the air, while the indoor unit is the evaporator and absorbs heat from the air. If the test room air temperature is 80 F, any outdoor temperature above 80 F can be simulated at the outdoor unit, while the indoor unit can control to any temperature below 80 F. If the test room is not air conditioned, operation on hot days is limited to high temperature cooling since the condensing unit can only raise its air temperature.

Recirculating air loops are relatively easy to implement if complete heat pumps are available. No additional heating or cooling sources are required. However, the heat pump always rejects a significant quantity of heat to the room since the heat discharged from the condenser coil always exceeds the heat absorbed by the evaporator coil (by an amount equal to the power input). In addition, a large part of the engine reject heat is discharged into the room. The room ventilation or air conditioning system must have sufficient capacity for the increased load.

With recirculating air loops, the heat pump must be able to perform defrost cycles in the heating mode.

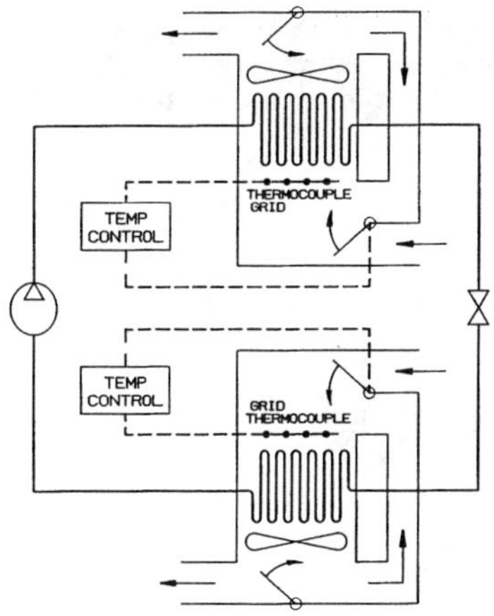

FIGURE 3. RECIRCULATING AIR LOOPS USED FOR LOAD CONTROL

One concern with this test method is that the fresh air entering the outdoor unit during heating will contain more moisture than normal, leading to excessive defrost cycles, particularly if the building is humidified in the winter. This problem is somewhat mitigated by the high percentage of recirculating air when simulating cold temperatures. During cooling operation, the return air humidity to the indoor coil is a concern. If a large percentage of the air is recirculated, the air can become excessively dry, reducing the heat exchanger effectiveness.

A summary of advantages and disadvantages of the recirculating air loop approach are:

Advantages

- Approach is easy to implement if complete heat pump is available
- No additional heating or cooling sources are required
- No need to alter heat pump design
- Defrost cycles are required

Disadvantages

- Temperature control is less accurate than liquid baths

- Potential for excessive defrosting
- Potential for air at evaporator coil to become excessively dry

Hot Gas Bypass Loop

The hot gas bypass loop is a simple loading approach that eliminates the evaporator coil and vastly reduces the heat rejected at the condenser coil. This approach minimizes the need for external heating and cooling devices. The operation is illustrated in Figure 4.

At the compressor outlet, a portion of the refrigerant bypasses the condenser through a throttling valve, while the remaining refrigerant passes through the condenser and expansion valve. The outlet of the bypass valve is a superheated vapor, while the outlet of the expansion valve is a two-phase mixture. The two paths come together in a mixer, with its outlet condition determined by the relative proportion of refrigerant from each valve.

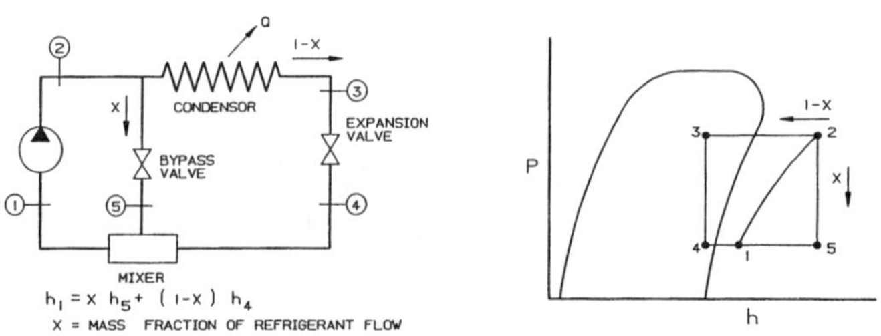

FIGURE 4. HOT GAS BYPASS LOOP

The condenser temperature can be controlled by using either a liquid bath or recirculating air loop. Either the liquid bath or recirculating air loop could be used. To simulate the desired evaporator temperature, the pressure difference across the bypass and expansion valves (operating in parallel) is adjusted until the suction pressure matches the vapor pressure for the desired evaporator temperature. The expansion valve must sense and control the degree of superheat at the compressor inlet.

To illustrate the operation of the hot gas bypass loop, assume we want to simulate the heating mode with a condenser coil temperature of 85 F and an evaporator coil temperature of 20 F (vapor pressure of 298 KPa for R22). The liquid bath or recirculation loop is first adjusted to achieve the desired condenser temperature. The bypass valve is then gradually closed until the compressor inlet pressure reaches 398 KPa. As the bypass valve closes, the expansion valve adjusts its opening to maintain the desired superheat.

The main disadvantage of this approach is the difficulty of controlling compressor inlet conditions and the inherent risk of slugging the compressor with liquid. Also, a standard thermostatic expansion valve may not be appropriate since it must handle two-phase flow at its inlet.

A summary of advantages and disadvantages of the hot gas bypass loop are:

Advantages

- Evaporator is eliminated and condenser size is minimized

Disadvantages

- Difficulty controlling compressor suction conditions
- Need to alter heat pump design
- Defrost cycles are not needed.

Selection of Loading Scheme

Table 1 summarizes the key features of each testing approach that were considered. The recirculating air loops were selected as the most cost-effective approach with good load control and the ability to test the complete heat pump as built.

TABLE 1.

	Component Testing	Relative Cost	Achievable Operating Hours/Year	Load Control
Field Testing	Complete System	Moderate/High	3000-5000	None
Laboratory Continuous Testing	Complete System	Moderate	8760	None
Engine Dynamometer Compressor Calorimeter	Engine and Compressor	High	8760	Limited
Environmental Chambers	Complete System	High	8760	Excellent
Liquid Baths	Engine and Compressor and Refrigerant Loop	Low	8760	Excellent
Recirculating Air Loops	Complete System	Low	8760	Very Good
Hot Gas Bypass	Engine and Compressor and Refrigerant Loop	Low	8760	Good

FLOW LOOP DESIGN AND CONSTRUCTION

In order to control the return air temperature to the outdoor and indoor units, an enclosure was built around each that allows controlled recirculation of a portion of the discharge air, as well as inlet for fresh air. Each structure is further described below.

Outdoor Unit Enclosure

A well insulated, plywood structure was fabricated to enclose the outdoor unit of the heat pump (see Figure 5). The enclosure was built indoors on a concrete floor which is covered with styrofoam sheeting to provide further insulation. The outdoor unit discharges air from its side into a separate compartment within the enclosure. Two sides of this compartment consist of louvers. One of the louvers reroutes the air back into the enclosure, while the other discharges the air from the enclosure. This louver system is controlled by a single linear actuator which opens one louver as it closes the other, offering a full range of air mixing, from complete discharge out of the enclosure to complete recirculation within the enclosure. The recirculated air is mixed with enough fresh test room air, at a separate vent in the main enclosure, to maintain atmospheric pressure at the outdoor unit coil. The minimum temperature that has been achieved in the outdoor enclosure using a 5-ton gas engine heat pump is -5 F. The maximum temperature that the outdoor unit is programmed to operate at is 97 F, although temperatures in excess of 125 F are easily achievable.

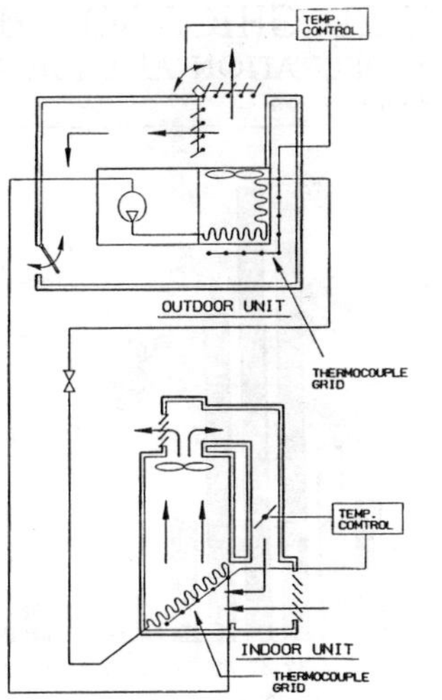

FIGURE 5. RECIRCULATING AIR LOOP CONFIGURATION

<u>Indoor Unit Enclosure</u>

For the indoor unit, a steel duct connects the discharge to its inlet. The flow through this duct is regulated by a computer-controlled damper, which regulates the amount of the recirculated air to be mixed with fresh air at the inlet. The inlet air temperature is kept at 80 F when the heat pump is operating in the cooling mode, and at 70 F in the heating mode.

<u>Data Acquisition System</u>

The gas heat pump was equipped with a Keithley Model 570 data acquisition and control system, which has 15 thermocouple inputs, 16 digital and 32 analog inputs, and 16 digital and 2 analog outputs. The input sensors are used to measure system temperatures and pressures, as well as fuel and refrigerant flows, from which GHP performance parameters are computed. All information is processed and stored on a computer. The control system was programmed to maintain evaporator and condenser return air temperatures by controlling linear actuators that regulate the position of air dampers at the indoor and outdoor units.

OPERATION AND LIMITATIONS

In actual operation there are several limitations in selection of operating mode and the temperatures to be simulated. When operating in the heating mode, the outdoor unit recirculates cold discharge air to mix with fresh air at room temperature. This produces a colder operating temperature at the outdoor unit, as desired. However, since the indoor unit recirculates warm air in the heating mode, the test room temperature must be maintained at or below the target temperature--70 F--for the indoor unit. Since the test room is not air conditioned, the heating mode cannot be run during the summer. To run in heating mode when the actual outdoor temperature is moderate requires good room ventilation to avoid raising the room temperature. Two large blowers were added for this purpose.

In the cooling mode the indoor unit serves as the evaporator. To maintain an evaporator inlet temperature of 80 F, the test room temperature must be maintained at 80 F or higher. This is generally not a problem, particularly in summer months, because the net effect of the heat pump is to raise the room temperature. On the other hand, the outdoor unit can only produce temperatures higher than room temperature. During the hottest times of the year, it can be difficult to keep the room temperature in the desired range. Again, good room ventilation is important to avoiding operating problems.

Other potential problems with the recirculating air loops stem from the lack of humidity control. During simulation of low temperature heating, a concern involved drawing in too much humid air to the outdoor unit, causing an excessive number of defrost cycles. This problem was avoided by simulating the heating mode only during colder weather, and by operating the units in a building without humidification Another concern, which arises during the cooling mode of operation, was operating the indoor unit with a dry coil. With a large amount of air recirculation at the indoor unit, return air humidity can become very low, thus reducing the evaporator coil efficiency and its temperature. However, by maintaining the test room temperature close to the desired indoor unit temperature, the amount of recirculated air is minimized preventing low return air humidity.

DUTY CYCLE OPERATION

An engine-driven gas heat pump is a very efficient space conditioning product, in part due to the inherent ability to modulate engine speed. As illustrated in Figure 6, the GHP essentially operates in three regimes. During moderate outdoor conditions, the unit operates at low speeds and cycles similar to conventional air conditioning equipment. As the need for heating or cooling increases, the GHP "modulates" its speed in order to satisfy a building's load requirements. At extreme outdoor conditions and building space conditioning load requirements, the system operates continuously at a maximum speed condition. During moderate heating conditions (20 to 40 F outdoor temperatures), the outdoor unit will also operate at maximum speed for short periods in a defrost mode whereby frost is melted on the outdoor coil by reversing the flow of refrigerant.

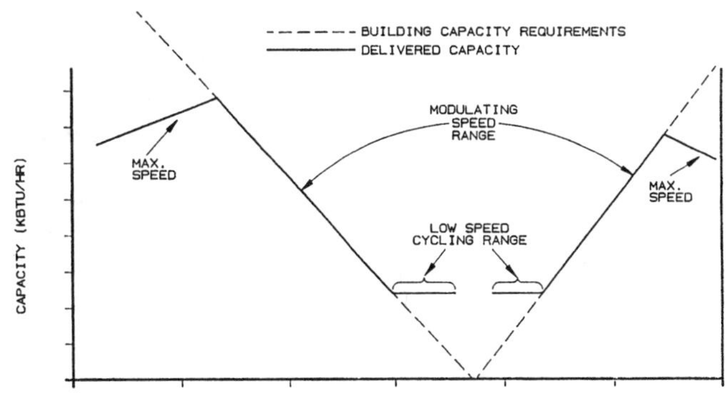

FIGURE 6. ENGINE DRIVEN GAS HEAT PUMP OPERATING REGIMES

In order to accurately assess the durability and reliability of the GHP and its key components--engine and compressor--a realistic operating schedule including number of operating hours, number of start/stop cycles, and average engine speed as functions of outdoor temperature was needed. Utilizing a computer seasonal performance analysis model to predict the operation and performance of the engine-driven GHP in icago and Salt Lake City, operating schedules were developed for two endurance units. Table 2 shows the complete operating schedule for Chicago, while Figure 7 shows the operating hours per year at various outdoor temperature for Chicago and Salt Lake City. The data acquisition and control systems were then programmed to control the operation of the GHP according to the appropriate duty cycles for each city. On a periodic basis, detailed inspections were performed on the engine, compressor, and other key components.

RESULTS AND CONCLUSIONS

The recirculating air loops have proved to be a viable method for performing continuous duty cycle testing of engine-driven GHPs. This method has allowed trouble-free, year-round, continuous operation simulating both Chicago and Salt Lake City operating conditions. System performance, including refrigerant mass flow, volumetric efficiency, capacity, and coefficient of performance (COP) has been measured during normal operation of the system. Measured compressor performance has demonstrated good agreement with compressor calorimeter test results. Monthly operating and maintenance costs have been minimal with more than 6,500 hours of operation achieved on each GHP over approximately 10 calendar months, including time for inspections and system startup.

Testing of a complete GHP system has eliminated the danger of testing separate components under laboratory breadboard test conditions not necessarily representative of the intended product design and enabled the evaluation of small components (e.g, belts, valves, etc.) not possible with independent component testing. Because operating conditions can be controlled with reasonable accuracy, this method also allows verification of GHP performance at any test condition.

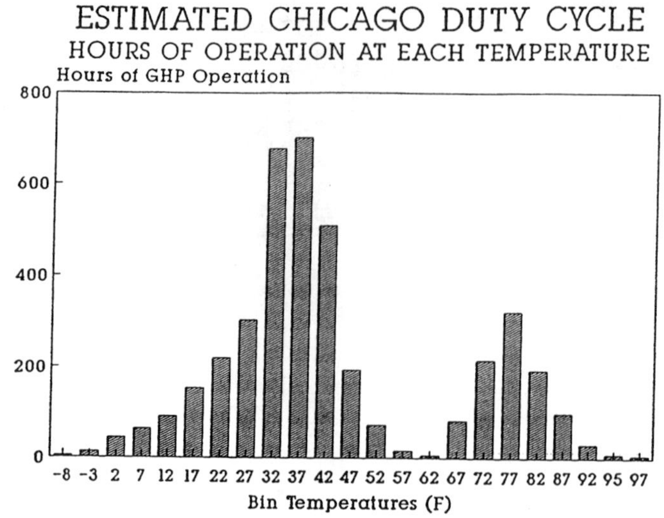

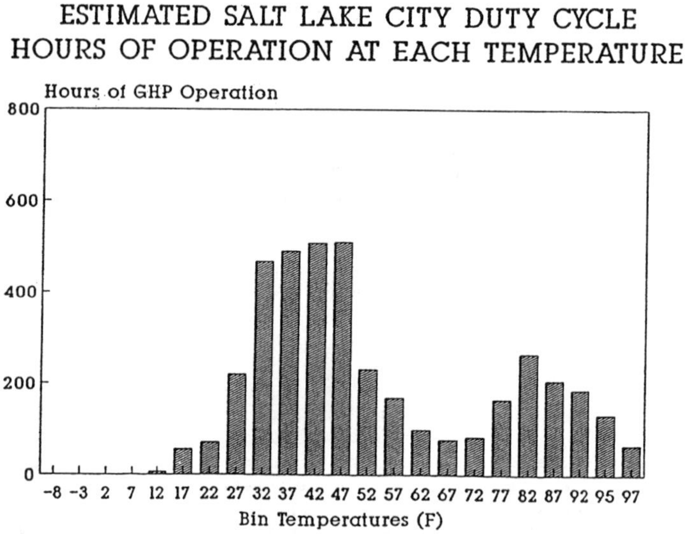

FIGURE 7. OPERATING HOURS-VERSUS-OUTDOOR TEMPERATURE

TABLE 2. ESTIMATED CHICAGO ANNUAL DUTY CYCLE

Operating Mode	Outdoor Temperature	Total Bin Hours	Operating Hours	Start/Stop Cycles	Engine (rpm)	Evaporator Temperature	Condenser Temperature
Cooling	97	8	6	0	2200	40	119
	95	12	9	0	2000	40	117
	92	39	30	0	1600	40	114
	87	129	99	0	1200	44	106
	82	251	193	0	1200	48	98
	77	454	321	166	1200	52	90
	72	579	215	658	1200	51	85
	67	543	82	342	1200	51	80
Heating	62	37	6	25	1200	47	94
	57	79	16	65	1200	42	92
	52	245	72	253	1200	37	91
	47	424	194	484	1200	32	89
	42	731	510	303	1200	27	88
	37	914	702	0	1200	22	91
	32	881	677	0	1200	17	94
	27	392	302	0	1500	12	95
	22	285	219	0	1900	7	95
	17	199	153	0	2400	2	93
	12	119	91	0	2800	-3	90
	7	82	63	0	2800	-8	88
	2	57	44	0	2800	-13	86
	-3	17	13	0	2800	-18	84
	-8	5	4	0	2800	-23	83

REFERENCES

(1) GRI Report No. 85-164, "Regional Characteristics and Heating/Cooling Requirements for Single-Family Detached Houses", Applied Management Sciences, August 1985.

(2) GRI Report No. 86-0235, "Gas Engine Heat Pump Cycle Analysis", Battelle Memorial Institute, October 1986.

(3) GRI Report No. 86-0192, "Phase 1-B Development of Kinematic Stirling/Rankine Commercial Gas Fired Heat Pump Systems", Stirling Power Systems, July 1986.

A NEW CORRELATION OF ADSORPTION EQUILIBRIUM

Lianquan Liu, Dhanes Charoensupaya, and Zalman Lavan
Department of Mechanical and Aerospace Engineering
Illinois Institute of Technology
Chicago, Illinois

ABSTRACT

In the analysis of adsorption cooling and heat pump systems, it is often found to be convenient to work with the isosteric diagram (plots of constant uptake). Several investigators have proposed linear isostere correlations based on experimental data of specific adsorption pairs. In this work, a new correlation of the adsorption equilibrium is derived from well established thermodynamic relations under the assumption that the latent heat ratio α is independent of temperature, and is expressed as $1/T_s = \alpha/T + \beta$. The parameters α and β are functions of uptake only. It is shown that with this new correlation the experimental work of equilibrium data measurements can be greatly reduced, and its simple form makes it easy to use in numerical simulations of adsorption cooling and heat pump systems. In the verification of the new correlation, experimental data of three widely used adsorbent-adsorbate pairs, namely silica gel-water, zeolite-water and activated carbon -methanol, are shown to be represented very well by the new correlation. Finally, Pons and Grenier's linear potential expression has been derived from the new equilibrium correlation using the relationship $\ln p_s = a + b/T$ for the vapor of the sorbate.

NOMENCLATURE

a, b	constants
A, B, C, D	parameters in equilibrium correlations, functions of uptake
L, L'	latent heat of vaporization, kJ/kg
p, p_s	pressure, saturation pressure of sorbate vapor, kPa
q_{st}	isosteric heat of adsorption, kJ/kg
R	gas constant, kJ/kg K
T, T_s	temperature, saturation temperature of sorbate vapor, K
w	uptake (mass of adsorbed phase per unit mass of adsorbent), kg/kg

Greek letters

α	latent heat ratio, $= q_{st}/L$
β	parameter, function of uptake, 1/K

INTRODUCTION

In the analysis of microporous adsorption such as the pairs of activated carbon-methanol, zeolite-water and silica gel-water, Dubinin-type equation (Bering et al., 1966, Dubinin and Astakhov 1971) is widely used to describe the equilibrium data. Ruthven (1984) pointed out the disadvantages of this theory. Masters and McEnaney (1983) observed deviations of the Dubinin-type equation from the experimental data and discussed the defects of the theory from a structural point of view. Pons and Grenier (1986) indicated that Dubinin's theory can not represent well the experimental data at small and large uptake. They pointed out that this discrepancy is due to intrinsic defects of the theory itself when saturation is approached, and due to the failure of the assumption the theory is based on at small uptake. Observing the linear behavior of the variation of the adsorption potential $\varepsilon = RT \ln(p_s/p)$ with temperature T, Pons and Grenier (1986) proposed a simple linear potential in the form $\varepsilon = A(w) + B(w)T$, and verified it using the measured data of activated carbon-methanol. A good agreement is achieved over the whole range of uptake, showing an improvement from Dubinin's theory. Hajji (1987) noticed the linearity of the isosteres in a $T - T_s$ plot for zeolite-water, and proposed an even simpler correlation in the form $T_s = C(w) + D(w)T$, with T_s being the saturation temperature of the sorbate and T the temperature of the sorbent-adsorbed phase. This form gives a great advantage in numerical analysis of closed cycle adsorption cooling/heat pump systems, since the working cycle of these systems can be most conveniently expressed in a $T_s - T$ diagram (Tchernev and Emerson, 1988), and its simple form is very easy to use. Hajji (1987) used this correlation in his numerical simulation of an adsorption cooling and heat pump system.

In spite of the advantages of the linear correlations mentioned above, they are of phenomenological nature and lack theoretical support. A serious restriction associated with the correlations of this kind is that they may only apply on a case-by-case basis, give no general conclusions for different adsorption pairs. In other words, they can not predict whether linear isosteres exist in general and under what condition if they do exit. By "linear isosteres" we mean any constant uptake curves which take the form of straight lines in an appropriate coordinate system. In this work, we will derive a new equilibrium correlation which answers these questions. We will aslo show that this new correlation agrees with the linear potential theory under certain condition.

DERIVATION OF THE NEW EQUILIBRIUM CORRELATION

Assume ideal gas behavior of the gas phase of adsorbate and ignore the volume of adsorbed phase in comparison of that of the gas phase. For an adsorbent-adsorbate system at an equilibrium state specified by temperature T and vapor partial pressure p, we have the widely-applied equation (Ruthven, 1984),

$$\left.\frac{\partial \ln p}{\partial T}\right|_w = \frac{q_{st}}{RT^2} \tag{1}$$

where q_{st} is the isosteric heat of adsorption, R the gas constant, and the subscript w indicates constant uptake.

By the chain rule of differentiation, equation (1) gives

$$\left.\frac{\partial \ln p}{\partial T_s}\right|_w \left.\frac{\partial T_s}{\partial T}\right|_w = \frac{q_{st}}{RT^2} \tag{2}$$

where T_s is the saturation vapor temperature corrosponding to pressure p. For saturated pure sorbate vapor Clapeyron equation gives

$$\left.\frac{\partial \ln p}{\partial T_s}\right|_w = \frac{L}{RT_s^2} \tag{3}$$

where L is the latent heat of vaporization. Substituting equation (3) into equation (2) yields

$$\left.\frac{\partial T_s}{\partial T}\right|_w = \frac{q_{st}}{L}\frac{T_s^2}{T^2} \tag{4}$$

Defining the latent heat ratio $\alpha = q_{st}/L$, seperating variables and integrating along w=constant, we get the following relation between T_s and T:

$$\frac{1}{T_s} = \frac{\alpha(w)}{T} + \beta(w) \tag{5}$$

An important assumption we have employed in obtaining Equation (5) is that the ratio $\alpha(w) = q_{st}/L$ is temperature independent, and further, is only a function of uptake w. Many authors assume the temperature independence of q_{st} itself, and obtain linear isosteres in LnP vs. $1/T$, which can be derived from Equation (1) directly (Ruthven, 1984). However, this leads to the result that the adsorbed phase has the same specific heat as that of the adsorbate vapor, and is strongly doubted (Pons and Grenier, 1986), because it is believed that the adsorbed phase should possess the similar properties as those of the sobate in bulk liquid phase, not in vapor phase. Othmer (1940) examined the behavior of the latent heat ratio L/L' for two different substances and concluded that the ratio will be substantially constant, especially for related compounds, over a much wider range than either of L or L', since the variation of both will follow the same law. Following Othmer's argument, we expect that the temperature independence of the ratio q_{st}/L shoule be a good assumption and it should apply over a wider range than q_{st} does. One way to check the validity of this assumption is by Othmer chart, which is based on the equation:

$$\left.\frac{\partial \ln P}{\partial \ln P_s}\right|_w = \frac{q_{st}}{L} \tag{6}$$

Linear isosteres in the $\ln p - \ln p_s$ plot indicate temperature independence of the ratio q_{st}/L. As an example of the justification of this assumption, Close and Banks (1972) showed this to be true for the silica gel-water by representing experimental data in Othmer chart. In Figure 1 we give the Othmer chart of hydrogen sulfide on H-mordenite zeolite as another example to support this assumption.

The reason to assume α to be a function of uptake is that the isosteric heat of adsorption q_{st} is generally a function of uptake while the latent heat of vaporization L is a property of pure sorbate only and has no connection with uptake.

VERIFICATIONS BY EXPERIMENTAL DATA

Experimental data of three adsorption pairs, which have been widely used in adsorption cooling/heat pump systems, are used to verify the new equilibrium relationship Equation (5). Figure 2 shows the data of silica gel-water, given by Charoensupaya (1986), in a $1/T_s - 1/T$ plot. Figure 3 gives the data from Pons and Grenier (1986) of activated carbon-methanol. Shown in Figure 4 are the data of Division® 3A molecular sieve-water vapor. Linear isosters are clearly seen in these figures and Equation (5) is thus justified. The parameters $\alpha(w)$ and $\beta(w)$ can be obtained in polynomials by curve fitting experimental data. Expressions obtained from the data in Figure 2 are given here as an example:

$$\alpha(w) = 1.9608 - 0.1345w + 8.2 \times 10^{-3} w^2 - 2.229 \times 10^{-4} w^3 + 2.286 \times 10^{-6} w^4$$

$$\beta(w) = -7.964 \times 10^{-4} + 3.672 \times 10^{-5} w + 7.013 \times 10^{-8} w^2$$

FURTHER DISCUSSIONS ABOUT THE NEW CORRELATION

Since the new correlation of adsorption equilibrium given by

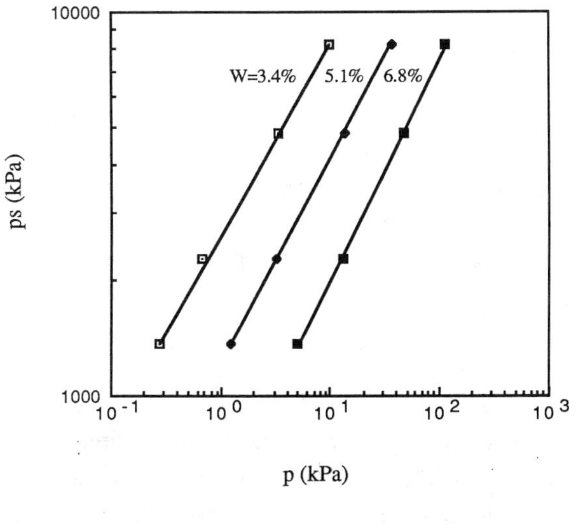

Fig. 1 Othmer chart of hydrogen sulfide on H-mordenite

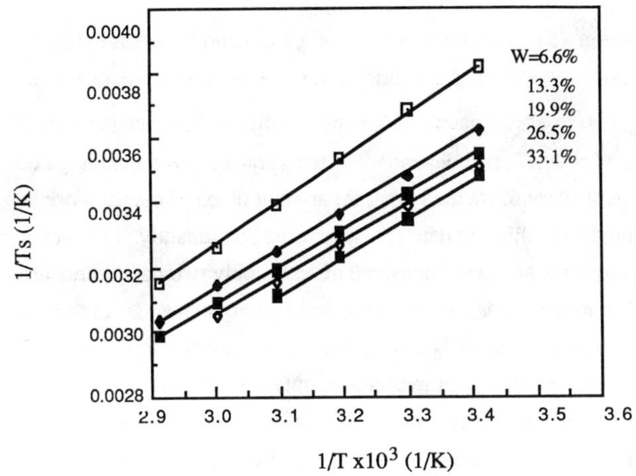

Fig. 2 Isosteres of Silica gel-water vapor, Charoensupaya's data (1986)

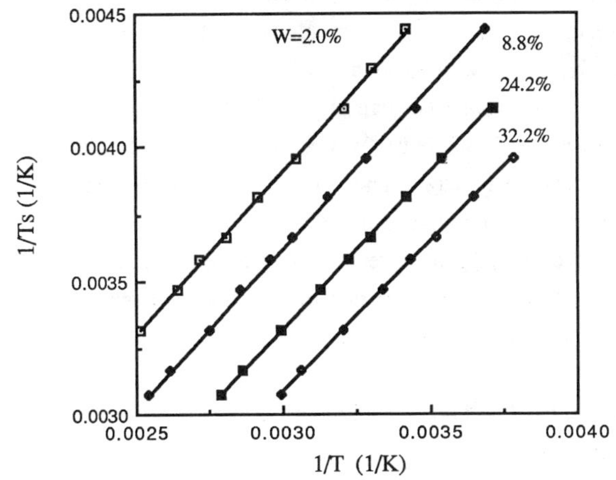

Fig. 3 Isosteres of activated carbon-methanol, Pons and Grenier's data (1986)

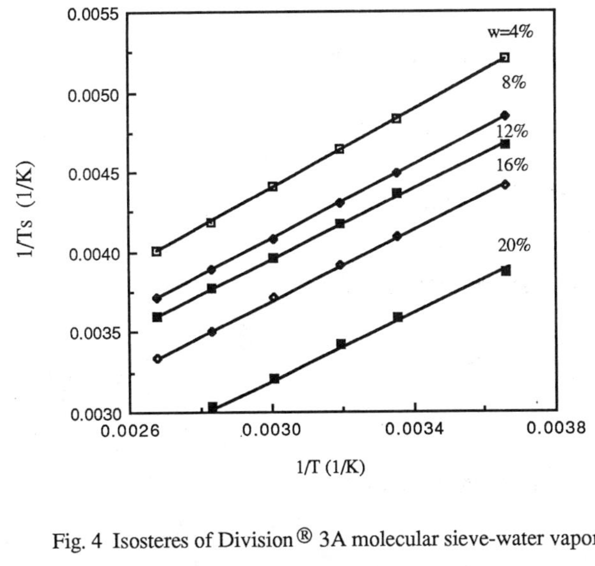

Fig. 4 Isosteres of Division® 3A molecular sieve-water vapor

Equation (5) is a direct result of established thermodynamic relations, it shows a general existance of linear isosteres in $1/T_s - 1/T$ coordinates, under the condition that the latent heat ratio $\alpha = q_{st}/L$ is independent of temperature. One advantage of this general relult is clearly that the amount of experimental work to obtain the equilibrium data can be reduced significantly. In the case that isosteres are to be measured directly, only two data points are needed to get one isostere because it is a straight line. If isotherms are measured first, only two isotherms are needed to get a set of isosteres over almost entire range of uptake except at the maximum w, (which is always the problem whenever isotherms are converted into isosteres). The process of converting isotherm data to isosteres is illustrated in Figure 5. Suppose two isotherms are given in $p - w$ coordinates. At any uptake w we have two data points (p_1, T_1) and (p_2, T_2). Using the pressure-temperature correlation of saturated sorbate vapor we can find the saturation temperature T_{s1} and T_{s2} corresponding to p_1 and p_2, respectively. Thus, two data points $(1/T_{s1}, 1/T_1)$ and $(1/T_{s2}, 1/T_2)$ are obtained and a liner isostere of uptake W is drawn. After a set of isosteres are found in this way, a curve fitting process is used to get the parameters $\alpha(w)$ and $\beta(w)$. By now the complete equilibrium correlation in the form of Equation (5) is obtained. As shown above, this approach requires less experimental data and therefore it either reduces the amount of experimental work or it can be useful when only a limited data are available.

REDUCTION TO THE LINEAR POTENTIAL CORRELATION

For many materials the dependence of the saturation vapor pressure on temperature can be expressed by

$$\ln p_s = a + \frac{b}{T} \quad (7)$$

where a and b are constants. This correlation has been applied to over 140 inorganic and organic compounds (Weast et al, 1988). A psychromatric chart in Figure 6 shows that this equation also reads

$$\ln p = a + \frac{b}{T_s} \quad (8)$$

Substituting equations (7) and (8) into (5) gives

$$RT \ln \frac{p_s}{p} = A(w) + B(w)T \quad (9)$$

where

$$A(w) = Rb(1 - \alpha(w)) \quad (10a)$$

and

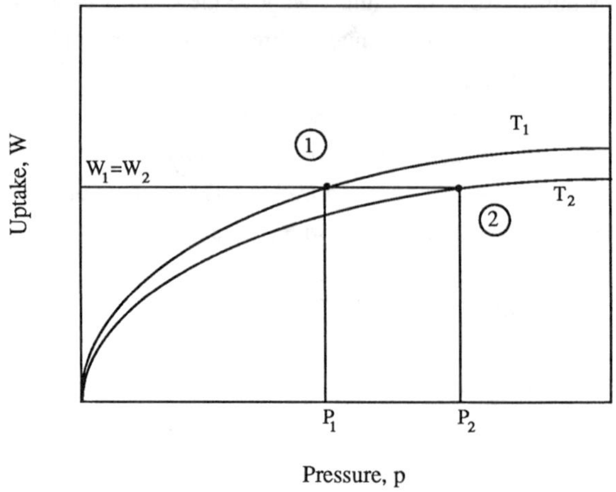

Fig. 5 Process of converting isotherms to isosteres

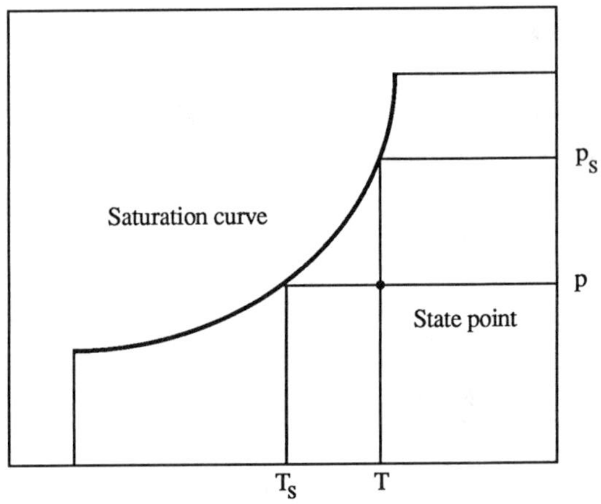

Fig. 6 Psychrometric chart showing the relations among T, T_s, p, and p_s

$$B(w) = -Rb\beta(w) \quad (10b)$$

Equation (9) is Pons and Grenier's (1986) linear potential equation. Therefore, although Pons and Grenier obtained their correlation as a phenomenological one, it could be a particular case of our linear correlation when the relationship Equation (7) applies. This reduction gives some theoretical support to the linear potential equation.

CONCLUSIONS

A new adsorption equilibrium correlation has been derived from Clapeyron equations under the assumption that the latent heat ratio q_{st}/L is temperature independent. This correlation shows that linear isosteres exist in the form of Equation (5) as long as the above assumption applies. The advantages of this new correlation include the reduction of the experimental measurements of the equilibrium data, and that the simple form of the linear relation of $1/T_s$ versus $1/T$ is easy to use in numerical simulation of adsorption systems, especially in the analysis of closed cycle adsorption cooling and heat pump systems.

The new relationship has been verified by experimental data of the pairs silica gel-water, zeolite-water and activated carbon-methanol. Good agreements between the new correlation and the data have been achieved.

It has been shown that Pons and Grenier's (1986) linear potential equation can be a particular case of the new correlation when the well-applied relation $\ln p_s = a + b/T$ is used for adsorbate vapor at saturation.

REFERENCES

Bering, B. P., Dubinin, M. M. and Serpinsky, V. V., 1966, "Theory of Volume Filling for Vapor Adsorption", *J. of Colloid and Interface Science,* Vol.21, pp. 378--393

Close., D. J. and Banks, P. J., 1972, "Coupled Equilibrium Heat and Single Adsorbate Transfer in Fluid Flow through a Porous Medium -- II: Predictions for a Silica-gel Air-Drier Using Characteristc Charts", *Chemical Engineering Science*, Vol.27, pp.1157--1169

Charoensupaya, D., 1986, "Experimental and Analytical Investigations of Composite Desiccant Structures and Low Humidity Adsorption", PhD thesis, Illinois Institute of Technology

Dubinin, M. M. and Astakhov, V. A., 1971, *Adv. Chem.* Vol.102, No.9

Hajji, A., 1987, "Analysis of Combined Heat and Mass Transfer in Closed-cycle Adsorption Cooling Systems", PhD thesis, Illinois Institute of Technology, Chicago, Illinois

Masters, K. J. and McEnaney, B., 1983, "Structural Analysis of Microporous Carbons Using the Dubinin-Radushokevich equation", *J. of Colloid and Interface Science*, Vol. 95, No. 2, pp. 340--345

Othmer, D. F., 1940, "Correlating Vapor Pressure and Latent Heat Data", *Industrial and Engineering Chemistry*, Vol. 32, No.6, pp.814--856

Pons, M. and Grenier, Ph., 1986, "A Phenomenological Adsorption Equilibrium Law Extracted from Experimental and Theoretical Considerations Applied to the Activated Carbon + Methanol Pair", *Carbon*, Vol. 24, No. 5, pp. 615--625

Ruthven, M., 1984, Principles of Adsorption and Adsorption Processes, John Wiley and Sons

Tchernev, D. I. and Emerson, D. T., 1988, "High-Effeciency Regenerative Zeolite Heat Pump", *ASHRAE Transactions*, Vol. 94(II), pp. 2024--2032

Weast, R. C. et al, ed. , 1988, Handbook of Chemistry and Physics, 69th ed., CRC Press

EFFECTIVE THERMAL CONDUCTIVITY IN WET DESICCANT BED WITH INERT MATERIALS

M. K. Sarwar and P. Majumdar
Department of Mechanical Engineering
Northern Illinois University
Dekalb, Illinois

ABSTRACT

This study presents the effective thermal conductivity of sorbing desiccant bed with inert material. The unit cell model is extended in order to account for the inert materials, and moisture sorbed in pores; and the effect of porosity, water content, and presence of inert material is determined. The prediction of the effective thermal conductivity from this proposed model is compared with that given by other available models and experimental data, and the proposed model is used to investigate the effect of porosity, water content and presence of inert material.

Nomenclature

g	Volume fraction of inert material in desiccant felt, m^3 of inert/m^3 of total materials
h	thickness of the unit skeleton, m
h_r	height of particle microroughness, m
k_A, k_g	thermal conductivity of air or gaseous phase in a pore, W/$m.K$
k'_A	total thermal conductivity in a micropore, W/$m.K$
k_c	contact thermal conductivity of granular material, W/$m.K$
k_{eff}	effective thermal conductivity, W/m.K
k_i	thermal conductivity of inert material, W/m.K
k_L	lower bound thermal conductivity, W/m.K
k_s	thermal conductivity of of solid phase, W/m.K
k_w	thermal conductivity of water, W/m.K
k_U	upper bound thermal conductivity, W/m.K
l	characteristic size of a pore
L	characteristic size of an elementary unit, m
m	thickness of water film, m
r_p	particle radius, m
R	thermal resistance, /kw
R_c	thermal resistance of solid contact between the two particles
R_{en}	thermal resistance of inert material
R_g	thermal resistance in gas phase
R_s	thermal resistance in solid phase
T	temperature, K
V_w	volume of water, m^3
V_p	volume of pore, m^3
V_s	volume of solid, m^3
V_{sw}	volume of solid plus water, m^3
w	water content kg H_2O/kg of dry gel
w_{max}	maximum amount of water that can be adsorbed by the microporous sorbent
x,y	space coordinates, m

Greek Letters Term

ρ	density of powdered material
ε	porosity
ε_f	felt porosity
ε_p	particle porosity
ρ_B	bulk density, kg/m^3
ρ_p	particle density, kg/m^3
τ	tortuosity factor

Introduction

In recent years, considerable research has focused on the use of solar energy or other low-temperature energy sources, and some of these studies have been concerned with the desiccant cooling system. A solid desiccant cooling system uses solid sorbents, such as silica gel, to adsorb moisture from humid air and is regenerated by using hot air from a solar air heater or other low-temperature air heater. A desiccant cooling system using adiabatic dehumidification has been investigated at IGT (1977), AiResearch (1978), Fauske and Associates (1982) and at SERI (1984). Also, a desiccant cooling system with a cooled fixed-bed desiccant dehumidifier was studied at IIT (1982). But these systems proved to be inadequate because they require a larger solar collector area or use costly natural gas. Investigators have proposed new methods for improving these systems. Advanced composite desiccant felt structures have been developed at IIT (1989). Epstein and Grolmes (1982) proposed to add inert, high-capacitance materials to the desiccant to improve the performance. Collier and Cohen (1988) proposed the concept of staged regeneration instead of inert heat capacitance to improve the performance of the system; they contend that this idea of stage regeneration is more effective than adding inert, high capacitance materials. The mathematical model used in these studies assumes a constant thermal conductivity for the sorbing media.

The objective of this study is to describe a model for effective thermal conductivity of a wet desiccant matrix structure with inert materials. This model will be used to study the effect of particle size, porosity, and percentage of moisture content on the effective thermal conductivity of the desiccant felt.

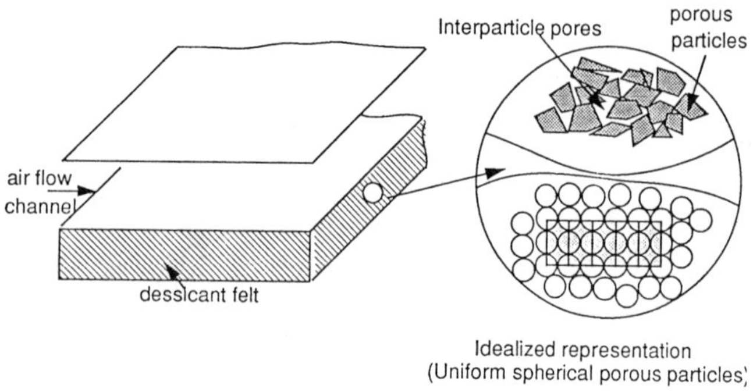

Figure 1.0 Schematic diagram of structure of desiccant

Effective Thermal Conductivity

The Schematic of a flow channel in a cooled bed dehumidifier is shown in figure 1. The wall separating the process and cooling channels includes a desiccant felt which is made of micron-size silica gel particles. Commercially available silica gel particles are fabricated into a thin felt of silica gel particles held in a teflon web which is itself bound to aluminum plates. The desiccant felt structure has interparticle macropores and intraparticle micropores. The general situation is shown in figure 1. Since the adsorption process is exothermic, heat is released as the desiccant adsorbs water vapor. The heat of sorption must be dissipated through the media which affects the sorption capacity of the materials, as well as outlet conditions of process air stream. The prediction of thermal energy dissipation in the porous desiccant felt must account for the two phases and composite nature of the system. The potential mechanisms for transport of energy in the felt can be identified as follows: (1) conduction through solid particles with pores, (2) conduction through the gas phase in the pore channels of the felt, (3) conduction across the liquid layer surrounding the points of contact between particles, (4) conduction through the points of contact between particles, and (5) vapor and liquid convection in the pores and microgaps at the place of contact caused by the temperature gradient.

The effective thermal conductivity, k_{eff}, is primarily dependent on the thermal conductivities of the composite phase, k_s, the liquid phase of water, k_w, and the vapor phase, k_A, and the structure of the porous felt and particles. The structure of the porous felt will not only affect heat and mass transfer processes, but also in turn will affect the relative position of the phases and the particles' contact points. Generally,the contact area will be higher for felts prepared at higher pressures, and it will depend on the amount of impurities, additives or inert materials present, as well as on particle shape and size distribution. Since the contact area is affected by many factors whose influence is not well established, the prediction of this area for the desiccant is very difficult. Various models have been suggested for effective conductivity of porous media. The analytical study of such systems using basic hydrodynamic and thermodynamic equations applicable to a continuous system is difficult. The simplest models take the overall value as an appropriate average of the separate phase contribution (Torquato, 1987), thus postulating that conduction through the solid acts over a solid volume, $1-\varepsilon$, and the conduction acts across the air in the voids and the liquid water layer around the points of contacts. An upper bound to the effective thermal conductivity is expressed by assuming an arithmetic mean corresponding to a parallel arrangement of the two phases as

$$k_{eff} = \varepsilon k_A + (1-\varepsilon)k_s \qquad (1)$$

The lower bound to the effective thermal conductivity is obtained by assuming a harmonic mean which corresponds to a series arrangement and is expressed as

$$k_{eff} = \frac{k_s k_A}{\varepsilon k_s + (1-\varepsilon)k_A} \qquad (2)$$

Woodside and Messmer (1961) proposed a weighted geometric mean of the effective thermal conductivity as

$$k_{eff} = k_s^{1-\varepsilon} k_A^{\varepsilon} \qquad (3)$$

Koh and Fortini (1973) presented the following equation which is similar to the equation for diffusion in a random pore structure.

$$k_e = \frac{k_s(1-\varepsilon)}{\tau} \qquad (4)$$

In attempting the formulation for regular and random packing, several investigators have suggested models using spheres, cylinders, cubes and unit elements and cells whose properties are assumed not to differ much from those of the system as a whole. Maxwell (1954) has established the effective conductivity for porous materials approximated by randomly distributed small spheres with the same volume fraction. Considering that the pores filled with gas of conductivity expressed as k_A (T) and that of the solid particles expressed as k_s(T), the overall effective thermal conductivity is approximated as

$$k_{eff} = \frac{2k_s + k_A - 2\varepsilon\left[k_s - k_A\right]}{2k_s + k_A + \varepsilon\left[k_s - k_A\right]} k_s \qquad (5)$$

Unit Cell Model

Luikov et al. (1968) have extended Dulnev's (1965) model of a unit cell to account for the finite contact area between the particles. The model is based on considering an elementary cell singled out from the porous structure, shown in figure 1 & 2, consisting of symmetrical particles and considering only quarters of two particles adjoining each other. The properties and mechanism of heat transfer for this cell is assumed to be similiar to that for the whole system. To establish the relationship between the porosity ε and the ratio h/l, consider only one quarter of the elementary cell, where h/2 is the thickness and width of the skeleton. Denoting V' as the total volume of the one quarter of the cell, V_1' as the volume of solid phase of the one-quarter of the cell, and V_2' as the volume of gaseous phase, following expressions can be shown:

$$V' = \frac{L^3}{4} \quad (6)$$

$$V_1' = (h/2)^2\,(2\,l + L) \quad (7)$$

$$\frac{V_1'}{V'} = 4\,\frac{(h/2)^2}{L^2}\,(3 - 2\,\frac{h}{L}) \quad (8)$$

Defining the volume fraction of the void as porosity, $\varepsilon = \frac{V_2'}{V'}$ and the volume fraction of solid phase as $(1 - \varepsilon) = \frac{V_1'}{V'}$, we can show

$$\frac{1-\varepsilon}{4} = \frac{(h/2)^2}{L^2}\,(3 - 4\,\frac{h/2}{L}) \quad (9)$$

Assuming $x = \frac{h/2}{L}$ and $\varepsilon_1 = \frac{1-\varepsilon}{4}$, equation (13) can be reduced to

$$4\,x^3 - 3\,x^2 + \varepsilon_1 = 0 \quad (10)$$

Also

$$\frac{h}{l} = \frac{h/L}{1 - h/L}, \quad \frac{h}{L} = \frac{h/l}{1 + h/l} \quad (11)$$

For a given porosity ε, ranging from zero to one, equation (10) is solved for $\frac{h/2}{L}$, and then equation (11) is solved for h/l. These results are curve fitted for this study to get the following expression:

$$\frac{h}{l} = 18.9234\;\varepsilon^4 - 56.0255\;\varepsilon^3 + 60.5757\;\varepsilon^2 - 30.0661\;\varepsilon + 6.6159 \quad (12)$$

Taking into consideration a network of series and parallel resistances as shown in figure 2, the effective thermal conductivity of the skeleton can be expressed as:

$$\frac{k_{eff}}{k_s} = \left[\frac{1}{\left[\frac{h}{L}\right]^2} + A\right]^{-1} + \frac{k_A}{k_s}\left[1 - \frac{h}{L}\right]^2 + 2\left[1 + \frac{h}{l} + \frac{k_s L}{k_A h}\right]^{-1} \quad (13)$$

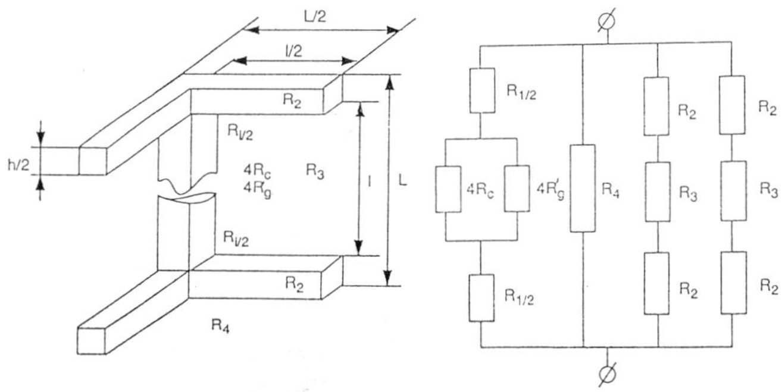

Figure 2.0 One quarter of elementary cell and its thermal resistances.

where

$$A = \frac{1}{\frac{k_c}{k_s} + \frac{v'}{4k_k k_m}\left[\frac{h}{L}\right]^2 10^3} \quad (14)$$

$$L = l + h$$

$$\frac{h}{L} = \frac{\frac{h}{l}}{1 + \frac{h}{l}}$$

$$v'_A = \frac{k'_A}{k_s}$$

k_k = particle adhesion coefficient

$k_m = \left[\frac{h_r}{L}\right] 10^{-3}$ = empirical coefficient

h_r = height of particle micro-roughness

Bjurstrom et al. (1984) carried out an experimental study of the thermal conductivity of beds of moist silica gel. The influence of porosity, water content, total gas pressure and temperature was determined through measurements under transient conditions with the transient hot strip (THS) method and under static conditions in a bench scale reactor.

A Model for Wet Dessicant Felt with Innert Materials

The unit cell model (Luikov et.al, 1968) is extended to account for the transmission of heat through water sorbed in the pores and through inert materials. It is assumed that adding of inert material will give an additional layer of desiccant of thickness $\frac{h}{2}(1-\sqrt{g})$ as shown in figure 3. For a volume fraction, g, of the inert material, the thickness of this layer will be $\sqrt{g}\frac{h}{2}$, where $\frac{h}{2}$ is the thickness of the total solid layer. In order to include the effect of moisture, it is assumed that the water adsorbed is distributed as a film with thickness m around the solid skeleton of the desiccant. If desiccant is assumed not to swell upon adsorbing water, the volume of water V_w in the pore, can be found by geometric calculations as:

$$V_w = V_{sw} - V_s \quad (15)$$

Where,

$$V_{sw} = \textit{Volume of solid plus water} = 4\,(h/2 + m)^2(l/2 - m + L) \quad (16)$$

and

$$V_s = \textit{Volume of solid} = 4\,h/2^2\,(l/2 + L) \quad (17)$$

Substituting equations (16) and (17) into equation (15) and expressing the volume of pore without any moisture as $V_p = l^2\,(L/4 + h/2)$, we can express volume V_w relative to V_p as:

$$\frac{V_w}{V_p} = \frac{m\left[3hl - 3hm + 3ml - 4m^2\right]}{l^2\left[\frac{L}{4} + \frac{h}{2}\right]} \quad (18)$$

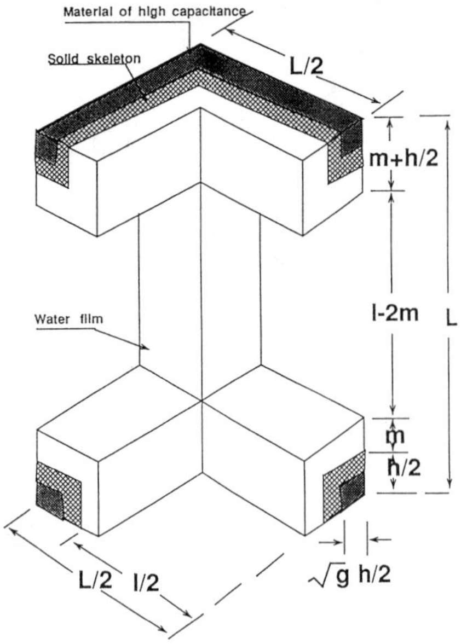

Figure 3 Model structure of a quarter of the unit cell with moisture and inert material.

and also as

$$\frac{V_w}{V_p} = \frac{w}{w_{max}} \tag{19}$$

where w_{max} is the maximum amount of water that can be sorbed by the sorbent. Equations (18) and (19) are combined and rearranged to express m/l as:

$$4\left(\frac{m}{l}\right)^3 - 3\left(1-\frac{h}{l}\right)\left(\frac{m}{l}\right)^2 - 3\frac{h}{l}\frac{m}{l} + \frac{w}{w_{max}}\frac{1+3\frac{h}{l}}{4} = 0 \tag{20}$$

Thermal resistances associated with this unit cell skeleton are shown in figure 4a and 4b. Figure 5 shows the thermal resistance diagram of the unit cell. In representing the thermal resistances, the first term of the subscript indicates the type of substance and second term indicates its thickness. Using the electrical analogy, the thermal resistances of the quarter of the unit cell are then written as

$$(R_{eq})^{-1} = (R_{en},L)^{-1} + (R_s,L)^{-1} + (R_w,L)^{-1} + 2(2R^1_{en},\sqrt{g}\frac{h}{2} + R^1_s,L-\sqrt{g}h)^{-1}$$

$$+\ 2(2R^2_{en},\sqrt{g}h/2 + 2R^2_s,\frac{h}{2}(1-\sqrt{g}) + R_w,l)^{-1}$$

$$+\ 2(2R^3_s,\frac{h}{2} + R^1_w,l)^{-1}$$

$$+\ 2(2R^3_{en},\sqrt{g}\frac{h}{2} + 2R^4_s,\frac{h}{2}(1-\sqrt{g}) + 2R^4_w,m + R^1_g,l-2m)^{-1}$$

$$+\ 2(2R^5_s,\frac{h}{2} + 2R^5_w,m + R^2_g,l-2m)^{-1}$$

$$+\ 2(2R^6_w,m+\frac{h}{2} + R^3_g,l-2m)^{-1}$$

$$+\ R_g,L^{-1} \tag{21}$$

Introducing the values of thermal resistances and dividing both sides of the expression by $\frac{k_s L}{4}$, we get the equivalent

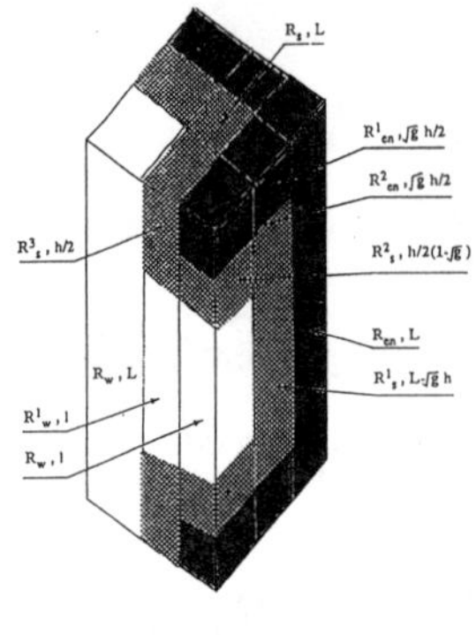

Figure 4b The thermal resistances of central column of quarter unit cell.

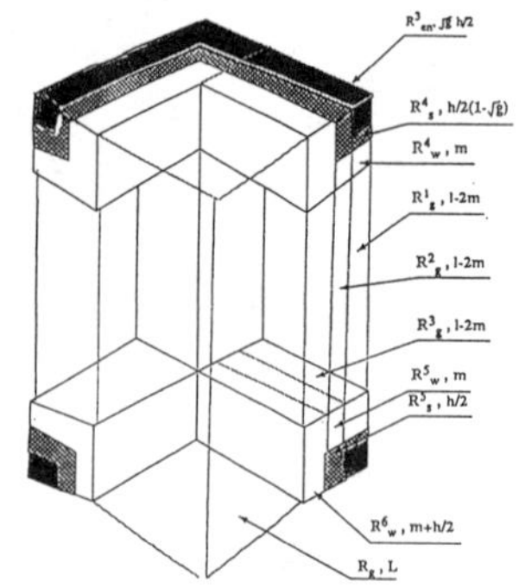

Figure 4a The resistances of different parts of the unit cell except for central column.

thermal conductivity of the whole particle unit cell as

$$\frac{k_{eff}}{k_s} = \frac{k_i}{k_s}\frac{gh^2}{L^2} + \frac{h^2\left(1-\sqrt{g}\right)^2}{L^2} + \frac{k_w}{k_s}\frac{4m^2}{L^2}$$

$$+2\left[\frac{k_s}{k_i}\frac{L}{h\left(1-\sqrt{g}\right)} + \frac{L\left(L-\sqrt{g}h\right)}{h^2\left(1-\sqrt{g}\right)\sqrt{g}}\right]^{-1}$$

$$+4\left[\frac{k_s L}{k_i m} + \frac{L\left(1-\sqrt{g}\right)}{m\sqrt{g}} + \frac{k_s}{k_w}\frac{Ll}{m\sqrt{g}h}\right]^{-1}$$

$$+4\left[\frac{L}{\left(1-\sqrt{g}\right)m} + \frac{k_s}{k_w}\frac{Ll}{mh\left(1-\sqrt{g}\right)}\right]^{-1}$$

$$+4\left[\frac{k_s}{k_i}\frac{L}{(\frac{l}{2}-m)} + \frac{L\,(1-\sqrt{g}\,)}{(\frac{l}{2}-m)\sqrt{g}} + \frac{k_s}{k_w}\frac{2Lm}{\sqrt{g}h\left(\frac{l}{2}-m\right)} + \frac{k_s}{k_g}\frac{2L}{\sqrt{g}h}\right]^{-1}$$

$$+4\left[\frac{L}{\left(1-\sqrt{g}\right)\left(\frac{l}{2}-m\right)} + \frac{k_s}{k_w}\frac{2Lm}{h\left(1-\sqrt{g}\right)\left(\frac{l}{2}-m\right)} + \frac{k_s}{k_g}\frac{2L}{h\left(1-\sqrt{g}\right)}\right]^{-1}$$

$$+4\left[\frac{k_s}{k_w}\frac{L\left(m+\frac{h}{2}\right)}{m\left(\frac{l}{2}-m\right)} + \frac{k_s}{k_g}\frac{L}{m}\right]^{-1}$$

$$+\frac{k_g}{k_s}\frac{4\left(\frac{l}{2}-m\right)^2}{L^2} \tag{22}$$

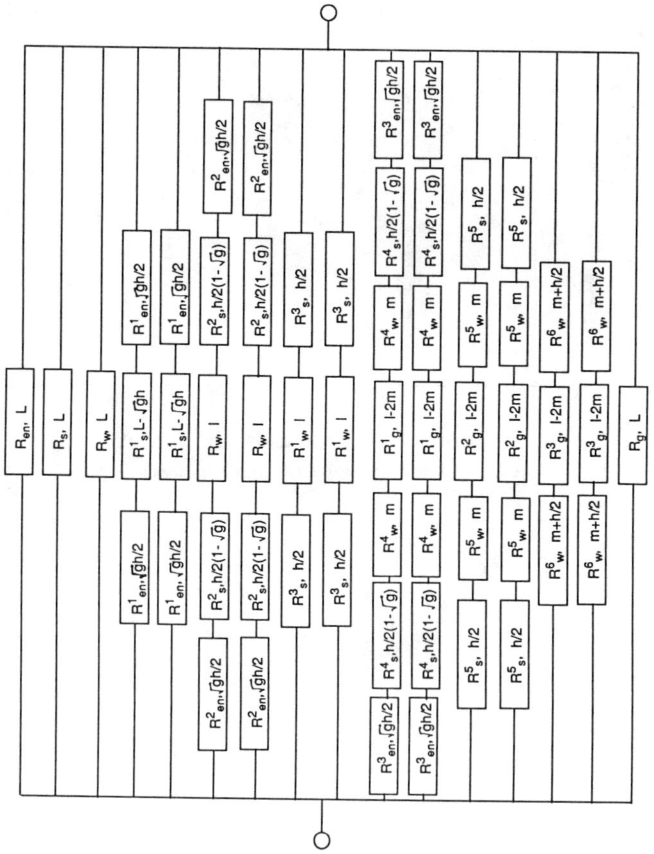

Figure 5 The thermal resistance diagram of unit cell having inert material and water layer of thickness m on it.

Results and Discussion

The effective thermal conductivity of the desiccant felt is calculated in two steps. First, the thermal conductivity of the particle is computed, for a given particle porosity, ε_p, pore diameter, and thermal conductivities, using equations (11), (12), (20), (22). Then, the thermal conductivity of the felt is computed, for the computed particle conductivity, felt porosity, ε_f, particle size, using same set of equations. Material data are presented in table 1. Thermal conductivity of dry desiccant felt is calculated using the proposed model; and comparison with other models, together with upper and lower bound predictions, are shown in figure 6. Results are also compared with the experimental data given by Bjurstrom (1984) for dry silica gel felt. Comparisons of these results show that the proposed model with negligible moisture content compares well with Luikov's model [13] and experimental data [15].

TABLE 1. MATERIAL DATA FOR SILICA GEL AND OTHER DATA USED IN THE MODEL

Density, Kg/m^3	1201
Microporosity of particles	0.45
Pore diameter, nm	2.1
Specific surface, m^3/g	750
Particle size, micron	5
Thermal conductivity of silica gel, W/mK	0.22
Thermal conductivity of water, W/mK	0.6118
Thermal conductivity of air, W/mK	0.0168

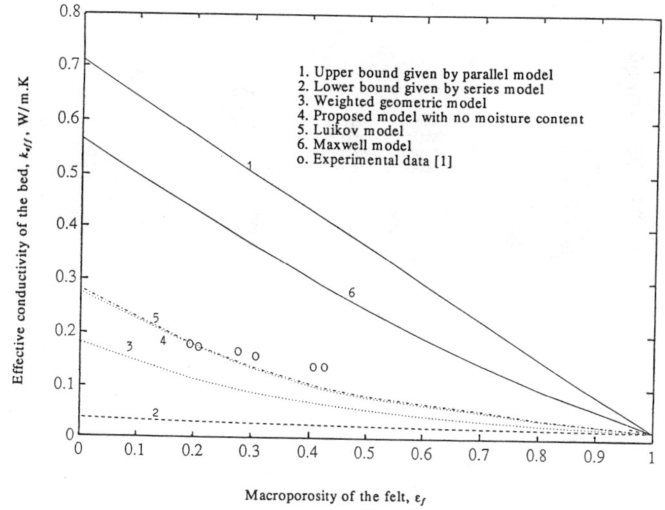

Figure 6. Comparison of Effective Thermal Conductivity Models

The model for moist felt with inert material is used to study the effective thermal conductiv ity of a desiccant felt with varying porosity, ε_f; percentage of moisture contents, W/W_{max}; and percentage of inert material, g. Results are presented in figures. Figures 7 and 8 show effective thermal conductivity with varying percentages of inert material for $\varepsilon_f = 0.197$ and $\varepsilon_f = 0.800$ respectively. The ratio of thermal conductivity of desiccant to inert material, k_i/k_s, is assumed to be 4. It can be seen that effective thermal conductivity increases, as expected, with an increase in moisture content and the fraction of inert material. However, the rate of increment is greater with a higher porosity. For felts with higher porosity, the conduction resistance due to the gas phase is very large compared to those with a lower porosity. With an increase in moisture content and inert material, there is a greater decline in gas-phase resistance, and this leads to a higher rate of increament of effective thermal conductivity. Figures 8 and 9 show the variation in effective thermal conductivity with var ying moisture content and fraction of inert material for $k_i/k_s = 4$, and $k_i/k_s = 6$ respectively. Effective thermal conductivity increases with the increase in conduction of inert material, and the effect is more important at a higher fraction of g.

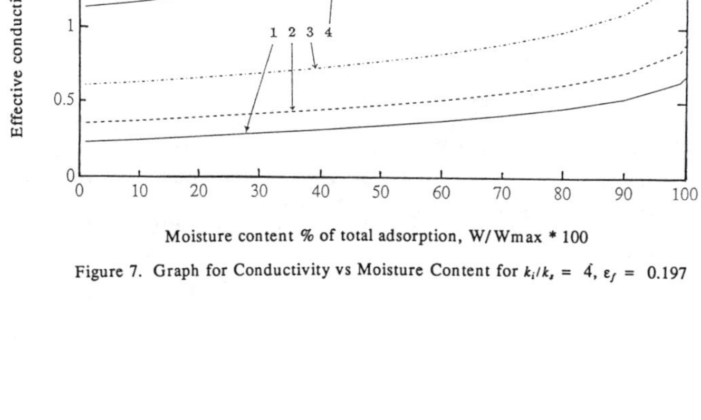

Figure 7. Graph for Conductivity vs Moisture Content for $k_i/k_s = 4$, $\varepsilon_f = 0.197$

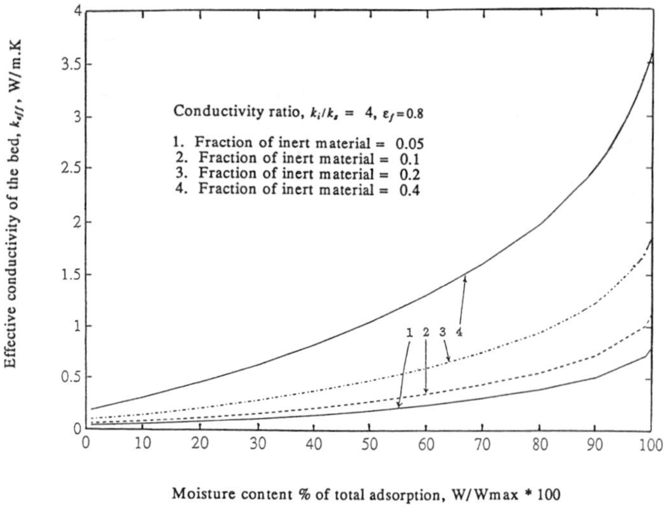

Figure 8. Graph for Conductivity vs Moisture Content for k_i/k_s = 4, ε_f = 0.8

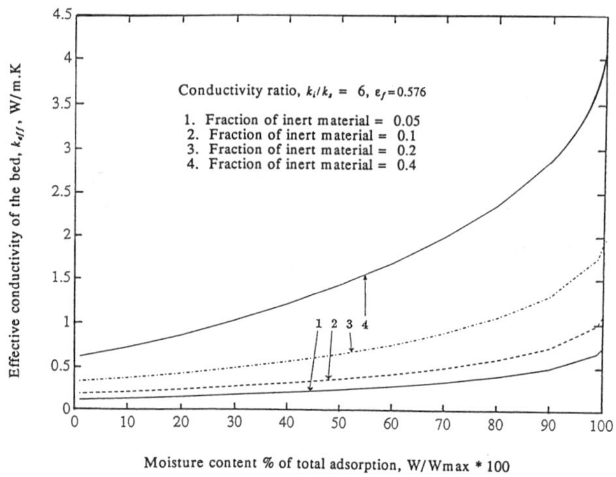

Figure 9. Graph for Conductivity vs Moisture Content for k_i/k_s = 6, ε_f = 0.576

Figures 10 through 13 show the variation of effective thermal conductivity with varying moisture content and fractions of inert material for g = 0.05, 0.1, 0.2, and 0.04 respectively. It can be noticed that at lower moisture content, effective thermal conductivity is high for lower porosity. As the moisture content increases, effective thermal condictivity also increases. However, the rate of increment is higher for higher porosity. With an increase in moisture content (about 80% for the case of g = 0.05), gas resistance decreases and effective thermal conductivity becomes higher for the case of higher porosity. It can also be seen that this cross-over takes place at lower moisture content with increases in fractions of inert materials. A similar trend is also noticed for the case of a higher ratio of thermal conductivity of inert to desiccant materials.

Conclusions

A model is developed for the calculation of the effective thermal conductivity of sorbing desiccant material with inert material. A paramatric study using this model shows the effects of porosity, moisture sorbed in pores, and presence of inert materials. It is shown that effective thermal conductivity of sorbing desiccant felt changes considerably with the porosity of the felt, size of the paricles, amount of moisture content, and fraction and thermal conductivity of the inert material.

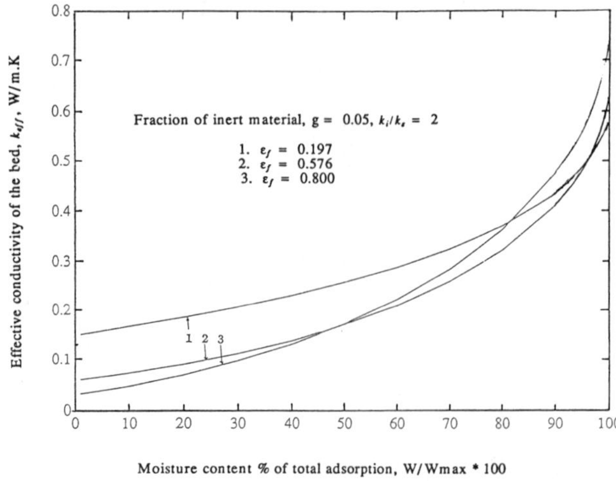

Figure 10. Graph for Conductivity vs Moisture Content for k_i/k_s = 2, and g = 0.05

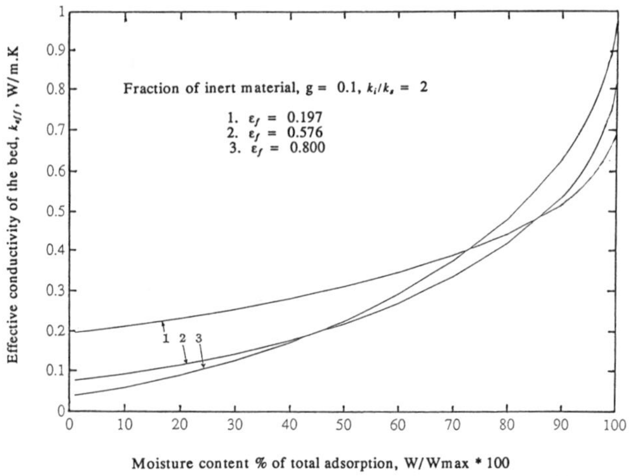

Figure 11. Graph for Conductivity vs Moisture Content for k_i/k_s = 2, and g = 0.1

References

1. Bjurstrom, H., Karawacki, E., and Carlsson, B., 1984, "Thermal Conductivity of a Microporous Particulate Medium: Moist Silica Gel." Int. J. of Heat and Mass Transfer, Vol. 27, pp. 2025-36.

2. Collier, R.K., and Cohen, B.M., 1988, "An Analytical Examination of Methods for Improving the Performance of Desiccant Cooling Systems." Solar Engineering Conference, Denver, pp. 139-144.

3. Dulnev, G.N., 1965, "Heat Transfer Through Solid Disperse Systems." INzh.-Fiz.Zh., 9(3), pp. 399-404.

4. Epstein, M., and Grolmes, M.A., 1982, "Desiccant Cooling System Performance: A Simple Approach." GRI-81/0089.

5. Koh, J.C.Y., and Fortini, A., 1973, "Prediction of Thermal Conductivity and Electrical Resistivity of Porous Metallic Materials." Int'l Journal of Heat and and Mass Transfer, Vol.16, pp. 2013-22.

6. Luikov, A.V., Shashkov, A.G., Vasiliev, L.L., and Fraiman, Y. E., 1968, "Thermal Conductivity of Porous Systems." Int. J. of Heat and mass Transfer, Vol. 11, pp. 117-40.

7. Majumdar, P., and Worek, W.M., 1989, "Performance of an open -cycle Desiccant Cooling System Using Advanced Desiccant Matrices." Heat Recovery Systems & CHP, Vol 9, No. 4, pp. 299-311.

8. Maxwell, J.C., 1954, A Treatise on Electricity and Magnetism, 3rd ed., Vol. 1, Chapter IX, Dover, New York.

9. Rousseau, J., and Hwang, K.C., 1978, AiResearch Report, The Third Workshop on the Use of Solar Energy for the cooling of Buildings, No. 78-14939.

10. Schlepp, D., and Barlow, R., 1984, "Performance of the SERI Parallel-Passage Dehumidifier.", NTIS SERI/TR-252-1951.

11. Staats, W.R., Wurm, J., Wright, L.R., Kunc, V., and Banasiuk, H.A., 1977, "Field Testing of Solar-MEC Systems Final Report." Institute of Gas Technology, Chicago, Illinois.

12. Torquato, S., 1987, "Bounds on the Thermal Conductivity of Disordered Heterogeneous Media." Thermal Engineering Joint Conference, pp. 359-67.

13. Woodside, W., and Messmer, J.H., 1961, "Thermal Conductivity of Porous Media." Journal of Applied Physics., Vol.32, pp. 1688-1706.

14. Worek, W.M., and Lavan, Z., 1982, "Performance of a Cross-cooled Desiccant Dehumidifier Prototype", Journal of Solar Energy Engineering, Vol.104, pp.187-196.

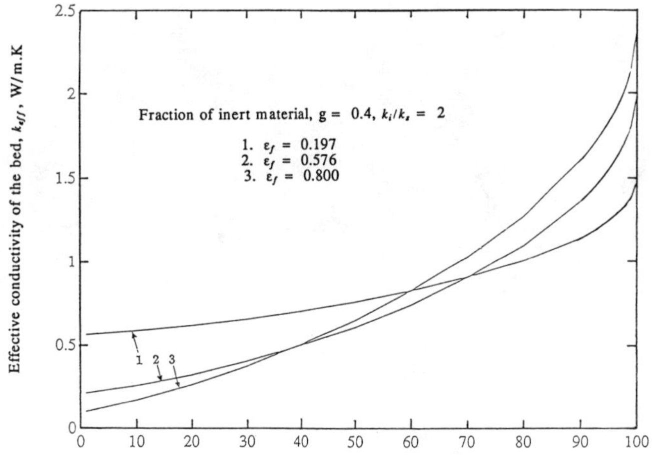

Figure 13. Graph for Conductivity vs Moisture Content for k_i/k_s = 2, and g = 0.4

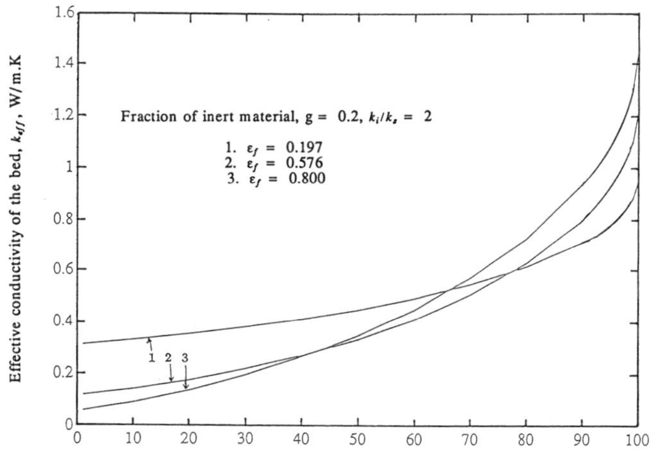

Figure 12. Graph for Conductivity vs Moisture Content for k_i/k_s = 2, and g = 0.2

COUPLED HEAT TRANSFER AND THERMODYNAMIC ADSORPTION HEAT PUMP ANALYSIS

Daniel J. Miles
Wave Air Corporation
Atlanta, Georgia

Sam V. Shelton
School of Mechanical Engineering
Georgia Institute of Technology
Atlanta, Georgia

ABSTRACT

Thermally driven adsorption heat pumps have been under increasing study for heat pump applications in residential, commercial, and industrial applications. These systems were first developed and commercialized as home refrigerators in the 1920's but essentially forgotten until the late 1970's. The efficiency of the basic cycle used in the 1920's is not high enough to be viable as a heat pump cycle against modern heat pump technology. Recent studies have shown that higher efficiencies require multiple adsorbent beds with heat regeneration between beds.

The adsorbent beds may consist of a fluid-to-solid heat exchanger. One possible configuration is a shell and tube type with fluid flow through the tubes and adsorbent packed in the shell. The solid adsorbent in each bed is alternately heated and cooled by the flow of hot and then cool heat transfer flow through the bed heat exchangers.The cooling of the adsorbent causes the refrigerant to be adsorbed at low pressure from the evaporator, while subsequent heating of the adsorbent causes the refrigerant to be desorbed. During the desorption process, the pressure in the bed first rises to the condenser pressure and then vapor is discharged to the condenser.

Past analyses show that regeneration of heat between two beds being heated and cooled 180 degrees out of phase can be accomplished efficiently and simply by the use of a heat transfer fluid loop between the beds in conjunction with a steep thermal profile in the beds along the fluid flow axis. Past analyses of these systems has used a transient, one dimensional heat transfer analysis coupled with a thermodynamic adsorption model. For ease of computation, these models have ignored any temperature gradients perpendicular to the direction of fluid flow; i.e., the adsorbent thickness around the fluid tubes is assumed small. This assumption does not allow optimization of the adsorbent layer thickness in order to minimize the heat exchanger size.

This current study includes a transient two dimensional heat transfer solution coupled with a thermodynamic adsorption model. Transient temperature profiles are calculated both along the axis of the tubes and in the adsorbent perpendicular to the tubes. This allows the study of varying tube spacing to minimize the bed size and weight per unit of heating or cooling capacity. Increasing the adsorbent thickness around each tube, (increasing tube spacing), increases the adsorbent per unit mass of tube; but at some point the adsorbent at the midpoint between tubes does not get fully heated and cooled, and therefore becomes ineffective.

This paper will present the coupled transient 2-D heat transfer and thermodynamic equations, discuss the numerical analysis used to obtain a solution, and present bed performance results for varying tube spacing (adsorbent thickness), and examine the bed size and weight per unit of cooling capacity.

NOMENCLATURE

A	area, m^2 (ft^2)
c	specific heat, cal/gm-C (Btu/lb-F)
h	fluid to tube heat transfer coefficient, W/m^2-K (Btu/hr-ft^2-F)
h_{AD}	heat of adsorption, heat of desorption, kJ/kg_{NH3} (Btu/lb_{NH3})
k	thermal conductivity, W/m-K (Btu/hr-ft-F)
L	tube length, m (ft)
ℓ	length coordinate, m (ft)
P	ammonia pressure, Pa (psia)
R	radius of carbon packed around each tube, m (ft)
r	radial coordinate, m (ft)
T	temperature, K (F)
t	time, hr
V	velocity, m/hr (ft/hr)

Greek Symbols

α	thermal diffusivity, m^2/hr (ft^2/hr)
γ	dimensionless tube temperature
θ	dimensionless adsorbent temperature
ρ	density, kg/m^3 (lbm/ft^3)
ϕ	dimensionless fluid temperature
Ψ	adsorption fraction, kg ammonia/kg zeolite (lbm/lbm)

Subscripts

A	adsorbent
C	condenser
E	evaporator
EFF	effective property of adsorbent/refrigerant pair
f	fluid
H	highest fluid temperature (heater outlet)
L	lowest fluid temperature (cooler outlet)
NH3	ammonia
t	tube

Superscripts

*	dimensionless quantity

INTRODUCTION

Thermally driven solid/vapor adsorption refrigeration was first demonstrated by Faraday in 1848. Commercial development of an adsorption refrigerator, or heat pump, occurred in the 1920's using silica gel and sulfur dioxide (Miller, 1929). This system was manufactured for residential and commercial food refrigeration. The same group made about 30 refrigerated railroad cars using the same technology with no electrical power required. Due to national electrification and the development of CFC refrigerants, the vapor compression cycle became the dominant refrigeration cycle, and the solid/vapor adsorption technology was basically abandoned until the late 1970's when it was considered for solar air conditioning (Tchernev, 1977; Meunier et al., 1979). Recent research interest has been focused on development of the cycle for use as a natural gas heat pump for space heating and cooling. As a natural gas driven heat pump, the cycle has the potential to reduce energy consumption, eliminate CFC and HCFC refrigerants, and prevent the need for new power plants to meet the increasing electric demand from electric space cooling equipment.

BASIC CYCLE DESCRIPTION

In the basic adsorption refrigeration/heat pump cycle, a condenser and evaporator are used providing the same function as in a vapor compression cycle, with a throttling device metering liquid refrigerant from the condenser into the evaporator. Two adsorbent beds are used to pull the low pressure vapor from the evaporator and discharge it to the high pressure condenser. These beds are each connected to both the condenser and evaporator with check valves allowing refrigerant to only flow in from the evaporator and flow out to the condenser. One adsorbent bed is heated while the second bed is cooled. The bed being heated desorbs refrigerant, pressurizing the bed. When this heated bed pressure reaches the condenser pressure, the refrigerant check valve opens allowing the refrigerant to flow to the condenser as the bed continues to be heated.

The cooling of the adsorbent bed causes it to adsorb the vapor in the bed. This depressurizes the bed until the pressure is reduced to the evaporator pressure, at which time the check valve from the evaporator opens to allow refrigerant flow to the bed as it continues to be cooled. When the first bed is fully heated and the second fully cooled, the process is reversed, cooling the first bed and heating the second. This returns the beds to their original state.

The efficiency of this basic cycle used in the 1920's is not high enough to be viable as a heat pump cycle against modern heat pump technology. However, there is an opportunity to improve the basic cycle efficiency. When one examines the basic cycle, it is noted that a substantial portion of the heat rejected from the bed being cooled is at a temperature equal to, or higher than, the temperature of the heat required by the bed being heated. Ideally this allows for substantial regeneration of that rejected heat to reduce the heater input.

REGENERATIVE CYCLE DESCRIPTION

Figure 1 shows a means for accomplishing this regeneration through the use of a heat transfer fluid loop consisting of four major components; i.e., the two beds, a fluid heater, and a fluid cooler. Initially, bed #1 is assumed cool and is to be heated, and bed #2 is assumed to be hot and is to be cooled. The fluid exits the heater at a constant high temperature, T_H, and enters bed #1. As it passes though this bed, the bed is heated while the fluid is cooled. The fluid exits bed #1 at a temperature near the initial bed temperature and enters the fluid cooler. The fluid is cooled further to a low temperature, T_L, and after passing through the pump and reversing valve it enters bed #2. As the fluid passes through this bed, the fluid picks up heat while cooling the bed. It then enters the heater at a high temperature close to the initial temperature of this bed. The fluid continues to circulate in this manner until bed #1 is sufficiently heated and bed #2 is sufficiently cooled. This completes the first half of the cycle.

The second half of the cycle is started by actuation of the heat transfer fluid reversing valve, which directs hot fluid from the heater to bed #2 and cool fluid to bed #1. This alternate heating and cooling of the two beds results in a near-continuous flow of refrigerant from the evaporator and to the condenser.

By proper bed heat exchanger design, a relatively steep temperature gradient can be established in the beds in the direction of the heat transfer fluid flow (Shelton et al., 1988). This "thermal wave" results in near complete heating (cooling) of the bed to a temperature T_H (T_L) while maintaining a low (high) temperature exiting the bed being heated (cooled). As the thermal wave moves through the heated (cooled) bed, the fluid outlet temperature starts to rise (fall). The normalized magnitude of this fluid temperature rise is the basis for cycle reversal. This "reversal criterion" is an operating parameter that can be varied to optimize performance. Note that the heat rejected in the fluid cooler can be utilized as useful heat along with the condenser heat rejection.

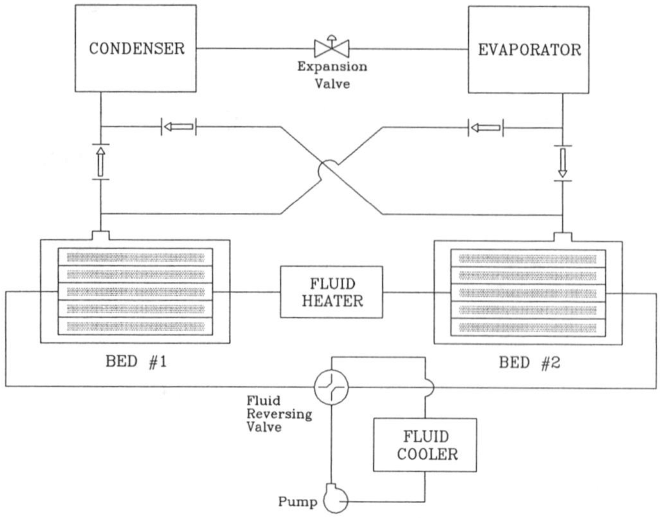

Figure 1. Solid adsorbent heat pump

PREVIOUS CYCLE ANALYSES

Prior to this study, the solid/vapor cycle analysis has consisted of a heat transfer model decoupled from the thermodynamic adsorption model. The thermodynamic adsorption analyses have used a one-dimensional thermal wave profile in the direction of the fluid flow simulated by a square wave (Shelton et al., 1989), as well as a more realistic ramp temperature profile (Shelton et al., 1990). The heating/cooling capacity and the coefficient-of-performance for the cycle was calculated. The effect of varying the steepness of the assumed thermal profile on the cycle performance was examined in these studies. Also, the ratio of the heat exchanger thermal mass to adsorbent thermal mass parameter was studied to determine the effect of this added non-adsorbent thermal mass on the cycle thermodynamic performance. Due to the one-dimensionality of the analysis, all temperature gradients normal to the direction of fluid flow were ignored, i.e., they were assumed uniform.

Each adsorbent bed is in effect a fluid-to-solid heat exchanger. One possible bed heat exchanger configuration is a shell-and-tube type with heat transfer fluid flowing through the tubes with adsorbent packed in the shell outside the tubes. A transient, one dimensional heat transfer analysis has been carried out for this shell and tube configuration, yielding a solution for the transient thermal wave temperature profile. Again, this analysis assumed that the temperature profile in the adsorbent normal to the tubes is uniform (Shelton et al., 1988). The transient one-dimensional thermal profile in the bed was calculated and described. This heat transfer analysis was decoupled from the thermodynamic model, and cycle performance was not calculated.

These transient one-dimensional analyses are only valid if the adsorbent thickness around the fluid tubes is assumed small. This requires closely spaced tubes, with very little adsorbent around each tube. The resulting high heat exchanger thermal mass which must be heated and cooled decreases cycle performance (Shelton et al., 1989). The high mass of heat exchanger metal per unit mass of adsorbent also increases the cost of the bed per unit of heating/cooling capacity.

CURRENT MODEL

These past one-dimensional decoupled analyses do not allow the calculation of the adsorbent temperature profile normal to the tubes, nor study of cycle performance with different tube spacing. The cycle parameters of interest are: (1) the cycle's thermodynamic performance and (2) the bed heat exchanger mass per unit of cycle heating/cooling output. In order to study and optimize this important tube spacing design parameter, a transient two-dimensional heat transfer analysis must be carried out and coupled with a thermodynamic adsorption analysis.

This study includes a transient two-dimensional heat transfer solution coupled with a thermodynamic adsorption model using zeolite as the adsorbent and ammonia as the refrigerant. Transient temperature profiles are calculated both along the axis of the tubes and in the adsorbent perpendicular to the tubes. Increasing the adsorbent thickness around each tube (increasing tube spacing), decreases the tube mass per unit mass of adsorbent; but at some point the adsorbent at the midpoint between tubes does not get fully heated and cooled by the tube and, therefore, becomes ineffective. This analysis studies the effect of varying this tube spacing on the cycle performance parameters.

Each bed can be visualized as a shell-and-tube heat exchanger with the tubes arranged in a square array as illustrated in Figure 2. The adsorbent cross-section associated with a single tube is represented by the shaded square area. For the purpose of this analysis, the square area is modelled as an equivalent circular area as illustrated. A general model of the transient heating and cooling of the solid adsorbent beds was developed based upon this configuration.

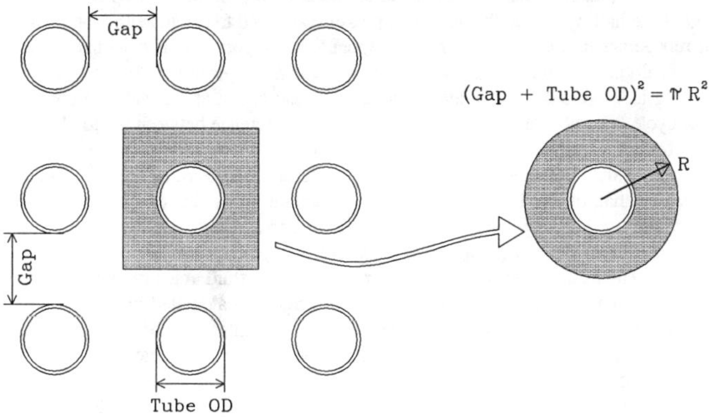

Figure 2. Adsorbent bed cross-section

Figure 3 illustrates the geometry and a characteristic temperature profile in the radial direction. The fluid and tube temperatures are assumed to be uniform in the radial direction and vary with respect to axial position, ℓ, and time, t. A finite heat transfer coefficient at the inside tube wall results in a temperature difference at the interface. The adsorbent temperature is equal to the tube temperature at $r=r_t$ and is bounded by an adiabatic surface at $r=R$. Thus the temperature in the adsorbent is dependent upon ℓ, t, and the radial coordinate, r.

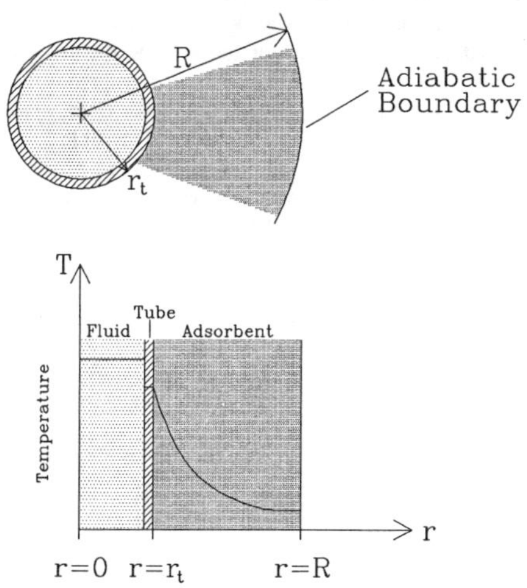

Figure 3. Characteristic radial temperature profile

SOLID/VAPOR ADSORPTION MODEL

The amount of ammonia adsorbed by zeolite is a function of the zeolite temperature and the ammonia pressure. A relationship between temperature, pressure, and the adsorption fraction, $\Psi=\Psi(T,P)$, for the zeolite/ammonia pair based on the Dubinin-Radushkevich equation from Critoph and Turner (1988) has been used in this analysis. The relationship between adsorption fraction and zeolite temperature at the evaporator and condenser pressures is illustrated in Figure 4.

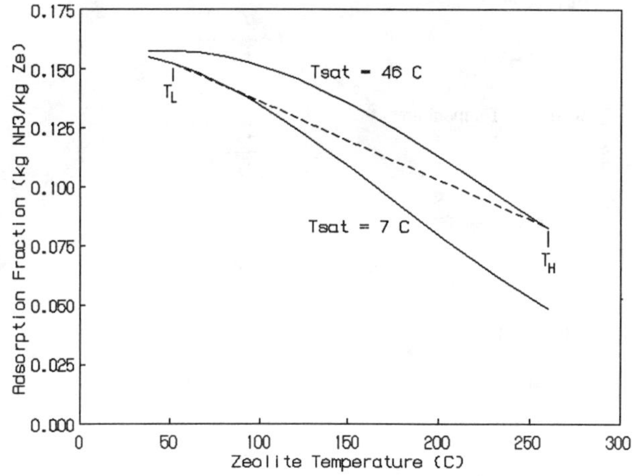

Figure 4. Zeolite/ammonia adsorption isobars

The adsorbent specific heat was modelled to include the effect of adsorbed ammonia and the heat of adsorption (desorption). The "effective" specific heat of the adsorbent is defined as:

$$c_{EFF} = c_A + c_{NH3}\left(\frac{\Psi(T_H,P_C) + \Psi(T_L,P_E)}{2}\right) + h_{AD}\frac{\partial \Psi}{\partial T}$$

The specific heat of the adsorbed refrigerant, c_{NH3}, was assumed to be equal to that of liquid ammonia at the average adsorbent temperature. Because the adsorption fraction varies during the cycle, an average of the minimum and maximum adsorption levels was used to calculate the contribution of the adsorbed refrigerant to c_{EFF}.

The heat of adsorption (desorption) is treated as a uniformly distributed heat source (sink) during cooling (heating) through the use of c_{EFF}. The strength of this heat source (sink) is dependent upon the sensitivity of refrigerant adsorption (desorption) to changes in temperature and the magnitude of the heat of adsorption (desorption). The sensitivity of adsorption (desorption) to temperature, $\partial\Psi/\partial T$, is approximated by the following:

$$\frac{\partial \Psi}{\partial T} \cong \frac{\Psi(T_L,P_E) - \Psi(T_H,P_C)}{T_H - T_L}$$

This corresponds to the slope of the dotted line in Figure 4. By use of the Clapeyron equation and the zeolite/ammonia adsorption data, the heat of refrigerant adsorption (desorption), h_{AD}, was found to be 2090 kJ/kg_{NH3} (900 Btu/lb_{NH3}).

ENERGY EQUATIONS

The equations governing this heat transfer process were developed using a control volume (CV) approach. The fluid, tube, and adsorbent were considered separately, resulting in three partial differential equations. The modes of energy transport considered in this analysis are the transport of energy into and out of the fluid CV associated with the flowing fluid, the transport of energy by axial conduction into and out of each CV, and energy transport between the fluid/tube and tube/adsorbent CV's by convection and conduction. The thermal capacity of the fluid, tube, and adsorbent are accounted for as is the heat of adsorption/desorption (treated as a uniformly distributed heat source/sink in the adsorbent).

An energy balance on each CV yields the three governing equations. The equations were nondimensionalized using the following dimensionless parameters:

Fluid Temperature:

$$\phi = \frac{T_f - T_L}{T_H - T_L}$$

Tube Temperature:

$$\gamma = \frac{T_t - T_L}{T_H - T_L}$$

Adsorbent Temperature:

$$\theta = \frac{T_A - T_L}{T_H - T_L}$$

Radial Position:

$$r^* = \frac{r}{r_t}$$

Tube Axial Position:

$$\ell^* = \frac{\ell}{r_t}$$

Time:

$$t^* = \frac{\alpha_{EFF} t}{r_t^2}$$

The resulting set of partial differential equations are:

Fluid Equation:

$$\frac{\partial \phi}{\partial t^*} = \frac{\alpha_f}{\alpha_{EFF}} \frac{\partial^2 \phi}{\partial \ell^{*2}} - \frac{r_t V_f}{\alpha_{EFF}} \frac{\partial \phi}{\partial \ell^*} - \frac{2\pi h r_f r_t^2}{\alpha_{EFF} \rho_f A_f c_f} (\phi - \gamma)$$

Tube Equation:

$$\frac{\partial \gamma}{\partial t^*} = \frac{\alpha_t}{\alpha_{EFF}} \frac{\partial^2 \gamma}{\partial \ell^{*2}} + \frac{2\pi h r_f r_t^2}{\alpha_{EFF} \rho_t A_t c_t} (\phi - \gamma) + \frac{2\pi r_t^2 k_A}{\alpha_{EFF} \rho_t A_t c_t} \left. \frac{\partial \theta}{\partial r^*} \right|_{r^* = 1}$$

Adsorbent Equation:

$$\frac{\partial \theta}{\partial t^*} = \frac{\partial^2 \theta}{\partial r^{*2}} + \frac{1}{r^*} \frac{\partial \theta}{\partial r^*} + \frac{\partial^2 \theta}{\partial \ell^{*2}}$$

The fluid, tube, and adsorbent were assumed to be at a uniform temperature at the beginning of the process. The model assumes that the bed is initially "cold" and that a heating half-cycle is undertaken first. This initial condition is expressed as:

$$\phi(\ell^*, t^* = 0) = \gamma(\ell^*, t^* = 0) = \theta(\ell^*, r^*, t^* = 0) = 0$$

Boundary conditions on the fluid, tube, and adsorbent temperatures were specified in accordance with the conceptual hardware configuration illustrated in Figure 1. An adiabatic boundary condition was applied to the tube and adsorbent temperatures at both ends of the bed (i.e. ℓ=0 and ℓ=L*) and to the adsorbent at r=R. It is understood that during a heating half- cycle, fluid flows in the direction from ℓ=0 to ℓ=L, and that fluid flows in the opposite direction during the cooling half-cycle. A constant fluid inlet temperature was used as the fluid inlet boundary condition, and an adiabatic boundary condition was applied to the fluid exit. The fluid conditions and their application differ for heating and cooling. Stated mathematically, the boundary conditions are:

$$\left. \frac{\partial \gamma}{\partial \ell^*} \right|_{\ell^* = 0} = \left. \frac{\partial \gamma}{\partial \ell^*} \right|_{\ell^* = L^*} = 0$$

$$\left. \frac{\partial \theta}{\partial \ell^*} \right|_{\ell^* = 0} = \left. \frac{\partial \theta}{\partial \ell^*} \right|_{\ell^* = L^*} = \left. \frac{\partial \theta}{\partial r^*} \right|_{r^* = R^*} = 0$$

Heating:

$$\phi(\ell^* = 0, t^*) = 1 \quad \text{and} \quad \left. \frac{\partial \phi}{\partial \ell^*} \right|_{\ell^* = L^*} = 0$$

Cooling:

$$\phi(\ell^* = L^*, t^*) = 0 \quad \text{and} \quad \left. \frac{\partial \phi}{\partial \ell^*} \right|_{\ell^* = 0} = 0$$

SOLUTION METHODOLOGY

In this analysis, the calculated temperature profile in the adsorbent at the end of the half-cycle and the ammonia pressure are used to calculate the total mass of refrigerant in the bed (adsorbed + vapor). The vapor mass is calculated from the ideal gas equation using the average adsorbent temperature. At the end of a heating half-cycle the bed attains condenser pressure, and at the end of a cooling half-cycle the bed is at evaporator pressure. The difference between the total mass of refrigerant at the end of the cooling and heating half-cycles is the amount of refrigerant that has been pumped to the condenser. (A zero or negative value indicates that the beds will not pump refrigerant from the evaporator to the condenser with the given set of input parameters.) The cooling capacity is calculated from the refrigerant mass flow and the cycle time.

The heater is assumed to deliver heat transfer fluid at a constant temperature T_H to the heated bed. Heat input to the heater is calculated by numerically integrating the product of the heat transfer fluid flowrate , specific heat, and the difference between T_H and T_f at $\ell^* = 0$ during the cooling half-cycle. The ratio of the cooling capacity to the heat input rate yields the cooling COP.

The governing set of three coupled partial differential equations were expressed in backward time, centered spatial finite difference form. The resulting implicit set of equations was solved numerically using an LU decomposition technique adapted from Press et al. (1986). The solution yields the fluid, tube, and adsorbent temperatures throughout the bed as functions of time. The axial and radial node spacing and time step size were sufficiently small to ensure accuracy without sacrificing computation time. This was verified by further reduction in spacing and time step size without significant change in the results.

RESULTS

The bed heat exchanger design which was analyzed consisted of 1.52 m (5 ft) long, 3.18 mm (0.125 in.) outside diameter steel tubes with a wall thickness of 0.25 mm (0.010 in.). The analysis was normalized on a per-tube basis, making the results independent of the number of tubes. The thermal mass of the heat exchanger shell was ignored, as well as any heat transfer down the shell in the axial direction. A commercially available heat transfer fluid, made up largely of diethylbenzene, was assumed with properties as listed in Table 1. Zeolite 13X was taken as the adsorbent with its physical properties shown also in Table 1, along with other details of the specific design. The gap between the tubes was varied from 0.25 mm to 20 mm (0.010 in. to 0.8 in.).

Table 1. Physical Properties and Operating Conditions

Physical Properties @ 150 C	Heat Transfer Fluid	Zeolite
Density (kg/m^3)	760	640
Specific Heat (cal/gm-C)	0.54	0.20
Thermal Conductivity (W/m-K)	0.10	0.14
Dynamic Viscosity (centipoise)	0.21	--

Operating Conditions:

T_H = 260 C
T_L = 52 C
T_C = 46 C
T_E = 7 C
Reversal Criterion = 0.4
Fluid Velocity = 12.2 m/hr

The resulting system performance parameters calculated are the system cooling COP and the cooling output per cycle per unit of zeolite mass.

The COP for varying tube gap is shown in Figure 5. This initially shows an increasing COP with an increasing gap. It reaches a peak and then declines with increasing gap. Figure 6 shows the refrigeration capacity per cycle per unit mass of zeolite for varying tube gap. It shows a continuously declining capacity with increasing gap.

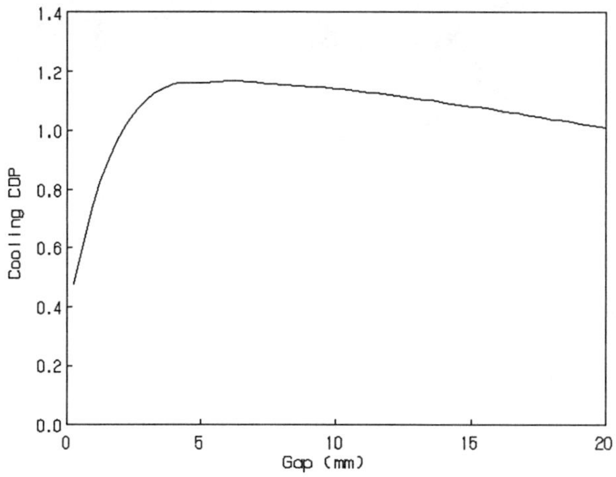

Figure 5. Performance variation with tube spacing

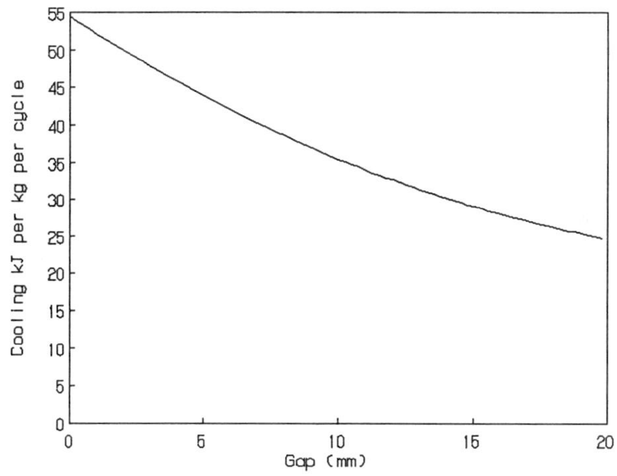

Figure 6. Cooling capacity variation with tube spacing

The decreasing capacity with increasing gap is due to the fact that as the depth of zeolite around the tube increases with the increasing gap, the temperature difference between the tube and the adsorbent at the center of the gap increases. This causes a decreasing utilization of zeolite, due to the fact that the temperature swing of the zeolite at the gap center decreases, thereby pumping less refrigerant. This dependence of the change in adsorption level on the change in adsorbent temperature can be seen from Figure 4.

The decreasing COP with increasing gap at large gaps seen in Figure 5 also follows from this ineffective utilization of adsorbent at the center of the large tube gaps. The adsorbent which is not fully heated and cooled to the tube temperature can act as a parasite on the system. For example, during the heating half-cycle, the pressure rises from the evaporator pressure to the condenser pressure. For adsorbent not fully heated, the increased pressure tends to cause an increase in the adsorption fraction. This pressure effect may be greater than the desired effect of decreasing adsorbent fraction with increasing temperature. Therefore, for very wide gaps where the adsorbent at the center of the gap undergoes very little temperature rise from the end of the cooling cycle to the end of the heating cycle, the adsorption level may actually rise rather than decrease. This means that refrigerant is pumped from the zeolite near the tube to the zeolite near the center of the tube gap rather than out to the condenser.

For an explanation of the decrease in COP with decreasing gap at small gap values seen in Figure 5, it is helpful to look at the tube mass per zeolite mass shown in Figure 7. Note that at the small gaps, the tube steel mass per unit of adsorbent mass rises exponentially. Therefore, in spite of the fact that small gaps lead to good utilization of the adsorbent due to small temperature drops perpendicular to the tube, a large faction of the heat put into the bed must go into heating the steel, with only a small faction going into heating the adsorbent. While a large fraction of this steel parasitic heat is regenerated, it still takes its toll on the COP at the very small gaps.

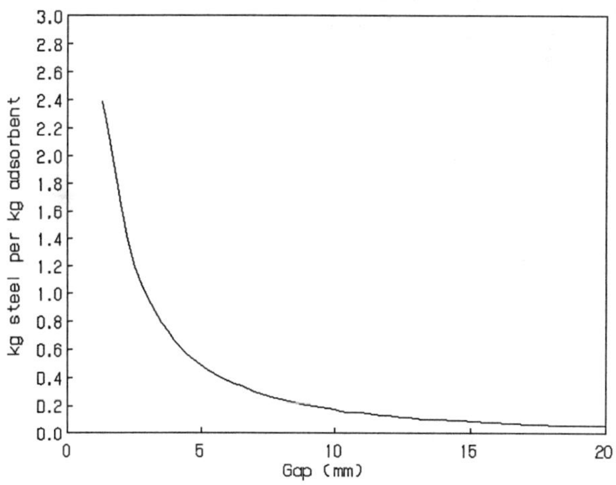

Figure 7. Mass ratio variation with tube spacing

CONCLUSIONS

The importance of an analysis of solid/vapor adsorption heat exchangers to include the calculation of the temperature profile normal to the adsorbent heating surface, which is the steel tube in this case, is paramount in modelling adsorption systems and optimizing bed heat exchanger designs. While simpler models, which assume this temperature profile to be uniform, can demonstrate the potential and general performance characteristics of adsorption refrigeration systems, it limits the usefulness of such models for predicting actual system performance, evaluating bed heat exchanger designs, and comparing heat exchanger configurations for cost effectiveness.

ACKNOWLEDGEMENT

This work was supported in part by the Gas Research Institute of Chicago, Illinois, under contract number 5085-242-1207.

REFERENCES

Critoph, R. E., Turner, H. L., 1988, "Activated Carbon-Ammonia Adsorption Cycle Heat Pumps", Proceedings of the CEC - British Gas International Workshop on Absorption Heat Pumps, April 1988, London, U.K.

Meunier, F., Mischler, B., Guilleminot, J. J., Simonot, M. H., 1979, "On the Use of Zeolite 13X-H_2O Intermittent Cycle for the Application to Solar Climatization of Buildings," SUN II, Proceedings of the International Congress of the Solar Engineering Society, Atlanta, GA, p. 739.

Miller, E. B., 1929, "The Development of Silica Gel Refrigeration," American Society of Refrigerating Engineers, Vol. 17, No. 4.

Press, W. H., Flannery, B. P., Teukolwky, S. A., Vetterling, W. T., 1986, Numerical Recipes, Cambridge Press.

Shelton, S. V.., Wepfer, W. J., and Miles, D. J., 1988, "External Fluid Heating of a Porous Bed," Chemical Engineering Communications, Vol. 71, p. 39.

Shelton, S. V., Wepfer, W. J., and Miles, D. J., 1989, "Square Wave Analysis of the Solid Vapor Heat Pump," Heat Recovery Systems & Combined Heat & Power, Vol. 9, No. 3, p. 233.

Shelton, S. V., Wepfer, W. J., and Miles, D. J., 1990, "Ramp Wave Analysis of the Solid/Vapor Heat Pump," ASME Journal of Energy Resources Technology, Vol. 112, No. 1.

Tchernev, D. I., 1977, "Exploration of Molecular Sieve Zeolites for Cooling of Buildings with Solar Energy," Final Report, NSF/RA-77017.

DESIGN OPTIMIZATION AND DEVELOPMENT OF AN ENERGY-EFFICIENT VAPOR COMPRESSION COOLING SYSTEM

M. P. Desai
Sundstrand Power Systems
San Diego, California

R. G. Mokadam, D. C. Nguyen, and F. L. Sindermann
Sundstrand Corporation
Rockford, Illinois

D. B. Wigmore
Sundstrand Power Systems
San Diego, California

ABSTRACT

The rapid technology development occurring in commercial vapor compression cycles reflects the growing concern over the greenhouse effect and ozone depletion issues relating to energy conversion and the chemical substances used in them.

The efficiency potential of the vapor cycle is receiving collateral interest in the aircraft community where its potential to significantly improve cooling efficiency over the currently used reverse Brayton air cycles will have significant benefits in payload/range and operating efficiencies as heat loads and fuel prices increase.

This paper describes the analytical design optimization of a new microprocessor-based, variable-speed, vapor compression cooling system which uses a binary nonazeotropic refrigerant mixture to provide a significant efficiency improvement to aircraft environmental control. Areas discussed include the thermodynamic and thermophysical modeling of the working fluid, cycle synthesis, and the performance-physical characteristic optimization process. Design tools developed for this effort are readily adaptable to heating/cooling systems for domestic and industrial application—space, ground-based, or other.

Although the system design was based on fully halogenated chlorofluorocarbon substances, subsequent work to identify blends and to quantify benefits of new working fluids that have acceptably low ozone depletion potential are reported upon. The paper also describes the status of system development scheduled to conclude with a flight test.

INTRODUCTION

A research and development program was started in 1981 to evaluate the technologies required for a new energy-efficient vapor compression cooling system. Apart from high design point performance, emphasis was also placed on efficient operation and precise temperature control over the wide spectrum of heat loads and heat sink conditions experienced in aircraft applications.

Analysis showed that a binary nonazeotropic refrigerant mixture (NARM) used in a patented thermodynamic cycle could provide significant performance improvements, while maintenance of high efficiency at off-design conditions could be realized by modulating the speed of the refrigerant compressor. Precise control was afforded using motor-driven expansion valves with speed, valve position, and other system functions controlled by a microprocessor. Performance, reliability, and logistics considerations led to the selection of a directly motor-driven hermetic compressor with refrigerant cooling and lubrication. Aerodynamic considerations stipulated that the 2-stage centrifugal compressor have the capability to run at 70,000 rpm. The provision of a condensate pump to provide refrigerant lubrication at startup allowed the cycle to be reconfigured by the microprocessor to perform as a pumped two-phase loop at low ambient temperature when compressor operation was not required.

Analytic design optimization required the development of a detailed model of the NARM, thermodynamic cycle, and physical characteristics of each system component. Principal among the physical models was the development of a numerical algorithm that accounted for the temperature profile of the evaporating or condensing NARM in the refrigerant heat exchangers. The independent states of this integrated system model were then manipulated to minimize a criterion function defined in terms of the weighted summation of power consumption and hardware weight. Thus, the interrelated effects of each refrigerant constituent, mixture ratio, and thermodynamic operating conditions of the cycle could be concurrently exercised to arrive at an optimum design.

In 1985, a system design was optimized around the avionics cooling requirements of a modern naval patrol aircraft and, with minimum modification, an advanced fighter aircraft. For the patrol aircraft application, evaporator heat load was 42.2 kW (12 ton) while suppling air at 7.2°C (45°F) and rejecting heat to air at 45.6°C (114°F), at sea-level static conditions. The best refrigerant was found to be 50 mole % R11/50 mole % R12. The predicted coefficient of performance (COP) was 1.6, and the flight weight was 495 lb (including evaporator and condenser fans).

From 1985 to 1989, parallel efforts to validate the theoretical models of the refrigerant, to develop the compressor, the power electronic motor drive, and the control electronics were followed by development testing of the complete system. These tests verified the integrity of the analysis and showed that handling, charging, and operating the system with a NARM was not very different from using a single substance.

Laboratory testing was completed in 1990 with an accumulation of over 250 hr of system operation. Preliminary performance measurements were made, and the seven digitally compensated control loops were closed such that push-button operation could be demonstrated in readiness for a 1991 flight test.

The international concern regarding potential destruction of the earth's ozone protection layer by migration of chlorofluorocarbon refrigerants to the stratosphere stimulated an examination of the capability of the emerging replacement materials. The ability of substitute fluids or mixtures of these substitute fluids to perform satisfactorily was studied. It was found, for some applications, that the designated replacement has significantly different thermodynamic properties and that a nonazeotropic mixture with its inherent performance advantages may be the only viable alternative.

Several studies have been made on the application of NARMs for cooling/heating systems (1, 2, 3, 4, 5, and 6). This paper will expand on the thermodynamic cycle and working fluid, analytic design techniques, and system development and alternative refrigerant studies that have been completed.

THERMODYNAMIC CONSIDERATIONS

Figure 1 shows a flow diagram of a vapor compression cycle using two heat or enthalpy exchangers: HX1 and HX2. HX1 is the "usual" unit for subcooling the high-pressure (saturated) liquid leaving the condenser by superheating the vapor leaving the evaporator before its entry into the compressor. This exchanger allows an increase in specific cooling effect (i.e., cooling effect per unit mass flow of the refrigerant), but the same result can be achieved if vapor superheating within the evaporator is permissible. The adverse effect of HX1 is an increase in the maximum cycle temperature T3. HX2 is a new element in the system. It is like another evaporator in parallel with the "main" evaporator. It provides subcooling of the condenser liquid from T5 to T6. Two expansion valves allow positive control of system operation.

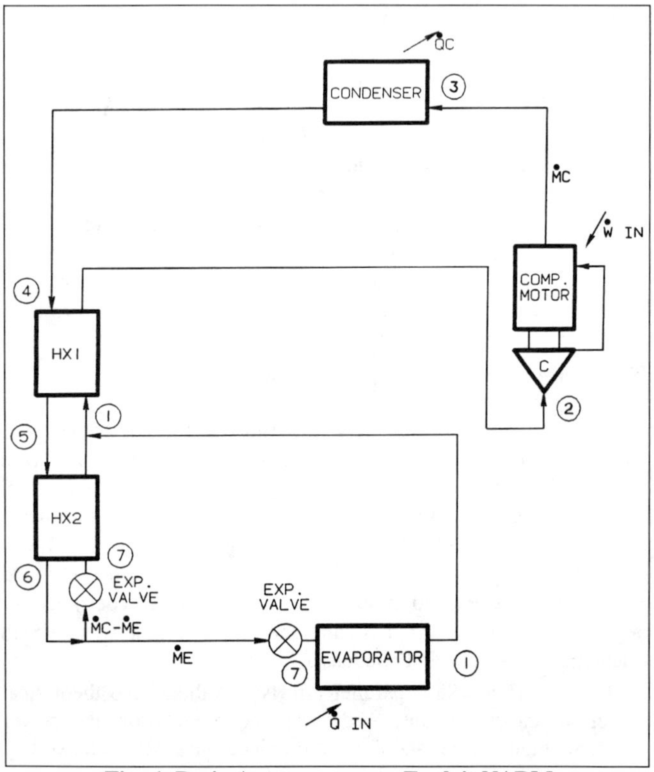

Fig. 1 Basic Arrangement to Exploit NARM

Figure 2 shows the cycle on a pressure-enthalpy (P-H) diagram. The saturation dome of the P-H diagram is qualitatively the same regardless of whether the working fluid is a single refrigerant or a binary mixture of a given fixed composition. As shown in this figure, the distinguishing feature is T7=T8=T1 for a single refrigerant and T7<T8<T1 for a NARM. For simplicity, pressure drops in the heat exchangers and connecting piping are neglected.

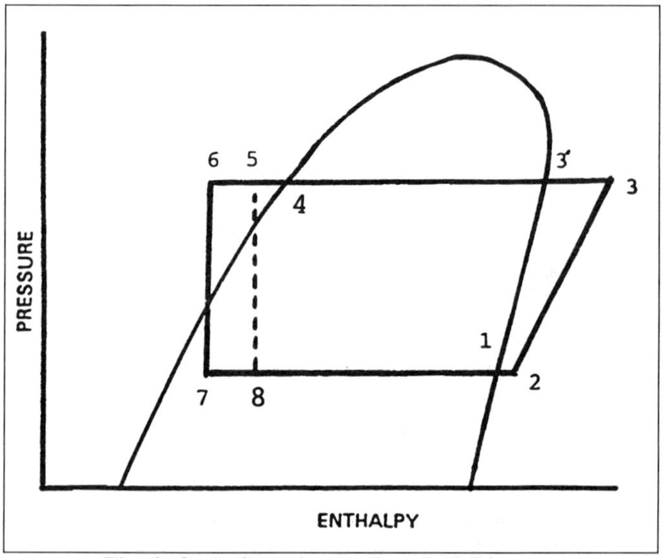

Fig. 2 Cycle Superimposed on P-H Diagram

Assuming steady state and neglecting potential and kinetic energy, we have:

For HX1

$$\dot{M}C\ (H2\text{-}H1) = \dot{M}C(H4\text{-}H5) \text{ or}$$
$$(H2\text{-}H1) = (H4\text{-}H5);\ T4>T2,\ T5>T1 \tag{1}$$

For HX2

$$\dot{M}C(H5\text{-}H6) = (\dot{M}C\text{-}\dot{M}E)\ (H1\text{-}H7) \text{ and}$$
$$H6 = H7;\ T1<T5;\ T6>T7 \tag{2}$$

Thus,

$$\dot{M}E/\dot{M}C = (H1\text{-}H5)/(H1\text{-}H6) = (H1\text{-}H5)/(H1\text{-}H7) \tag{3}$$

Where $\dot{M}$, H, and T represent the mass flow rate, enthalpy, and temperature of the refrigerant, respectively, and C and E refer to the condenser and evaporator.

The compressor work input $\dot{W}IN$ is given by:

$$\dot{W}IN = \dot{M}C(H3\text{-}H2) \tag{4}$$

Where H3 is obtained from the compressor/motor efficiency.

The cooling effect $\dot{Q}IN$ is given by:

$$\dot{Q}IN = \dot{M}C(H1\text{-}H6) = \dot{M}E(H1\text{-}H7) \tag{5}$$

From equations (3) and (5),

$$\dot{Q}IN = \dot{M}C(H1\text{-}H5) \quad (6)$$

From equations (4) and (6), the coefficient of performance, COP, is given by:

$$COP = \dot{Q}IN/\dot{W}IN = (H1\text{-}H5)/(H3\text{-}H2) \quad (7)$$

Note that HX2 affects the cycle in the following way.

1. The COP does not directly depend on HX2.
2. Without HX2, the fluid enters the evaporator at T8 rather than at T7. Because T7=T8=T1 for a single fluid, HX2 has no effect on the pressure ratio and the refrigerant temperature in the evaporator. In other words, HX2 is useless in a system using a single refrigerant.
3. For a NARM with T7<T8<T1, HX2 results in a reduction in temperature (from T8 to T7) at the evaporator entry for the same pressure ratio and COP as without it. Alternatively, a decrease in the pressure ratio and increase in the COP are obtained by using HX2 for an evaporator inlet temperature of T8. Thus, HX2 is essential for the efficient use of a NARM.

For a single fluid, heat is rejected from the cycle in the two-phase region at a constant T4. For a NARM, this process occurs over a temperature range from T4 to T3, which provides a benefit of lower cooling flow or lower condenser size.

As stated earlier, HX1 is not necessary if superheating can be accommodated in the evaporator. When two stages of compression are required, another heat exchanger HX3 can be interposed between the condenser and HX2 (or HX1, if used) to reduce compressor work and minimize cycle temperature. This is shown in Figure 3 for a cycle not requiring the suction-superheating function of HX1.

The cooling stream for HX3 is throttled to the interstage pressure level. It can come from the condenser or from the high-pressure liquid stream of any of the heat exchangers. The intermediate pressure fluid stream leaving HX3 is used to cool the compressor and its drive system before it is mixed with the first-stage exiting vapor stream. For details of the cycle and its variations refer to (7).

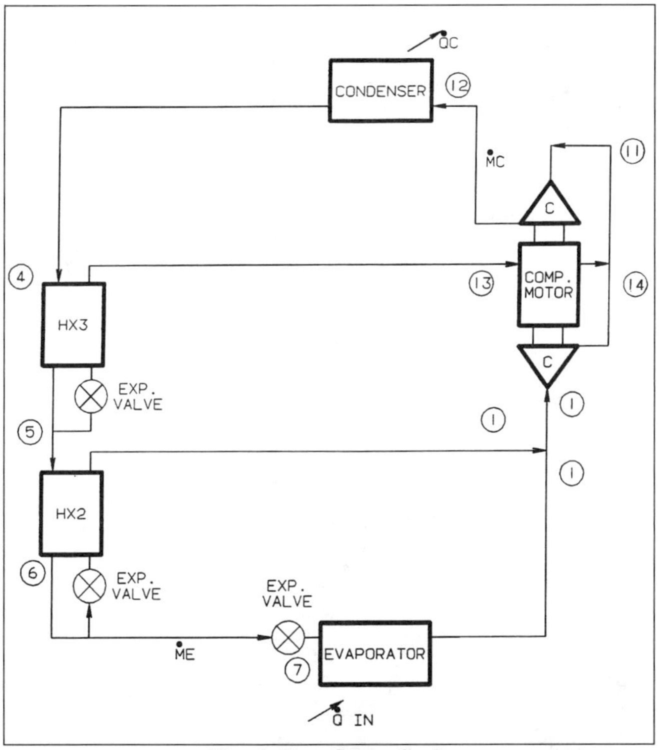

Fig. 3 Two-Stage Cycle

A NARM thermodynamic properties program based on a modified R-K-S equation (8, 9) was developed in the early 1980s to support design studies. For transport properties, we developed a computer program based on the models for polar fluids described in the text by Reid, Prausnitz, and Sherwood (10). The program enables evaluation of the thermal conductivity, viscosity, and diffusion coefficients of NARM in the liquid and vapor phases. In addition, a Sundstrand-funded program at the National Institute of Standards and Technology (NIST) produced a computer program for calculating thermal conductivity and viscosities of NARMs. This program is based on the NIST TRAPP (Transport Properties Prediction Program) routine developed for petroleum products. Although these models predate the NIST thermodynamic model (11), our experience shows them to satisfactorily describe the performance of experimental systems built to date.

PHYSICAL IMPLEMENTATION

Implementation of the two-stage NARM cycle is shown schematically in Figure 4, with the design point cycle shown on the NARM properties diagram in Figure 5. The heat exchangers are of plate-fin brazed aluminum construction arranged in a counterflow configuration to exploit the NARM two-phase temperature gradient. A two-stage centrifugal compressor directly driven by a permanent magnet brushless dc motor, supported by refrigerant liquid tilting pad hydrodynamic bearings, was designed to form a hermetic assembly. The compressor, which was designed to run at speeds of up to 70,000 rpm, is powered by a solid-state pulse-width modulated (PWM) inverter from conditioned aircraft power. The compressor surge valve and expansion valves are stepper motor-driven poppet types. A low suction head capable pump was introduced downstream of HX3 to provide bearing lubrication and motor cooling flow at startup and to circulate refrigerant at low ambient temperature when compressor operation is not required. Start/stop and mode sequencing and system control are under the authority of the microprocessor.

Liquid refrigerant leaving the condenser enters HX3 where it is further subcooled by evaporating the two-phase refrigerant at approximately interstage pressure. Because the NARM evaporates with a rising temperature, the flow for this two-phase cooling is tapped from the subcooled liquid leaving the subcooler. The two-phase NARM leaving the subcooler provides further two-phase cooling for the motor drive (speed controller) and compressor motor back iron cooling passages. A small portion of this two-phase flow is injected into the motor cavity for internal motor cooling. The rest of the two-phase flow is injected into the superheated vapor leaving the first compressor stage where the remaining liquid is evaporated to reduce the temperature of the superheated vapor entering the second-stage compressor.

The subcooled liquid leaving HX3 enters HX2 for further subcooling and then is routed to the compressor for bearing lubrication and motor gap cooling. The subcooled liquid flow leaving the bearings is then routed to the evaporator, a small portion being used for two-phase cooling in HX2. The outlet of HX2 is merged with the outlet of the evaporator to provide a minimal vapor superheat at the inlet of the first-stage compressor. To complete the cycle, the working fluid is subject to two stages of compression after which heat is rejected from the cycle to ambient air by desuperheating and condensing the refrigerant back to HX3 entry conditions.

The system control strategy is as follows. The compressor speed is modulated to satisfy the minimal superheat requirement at the

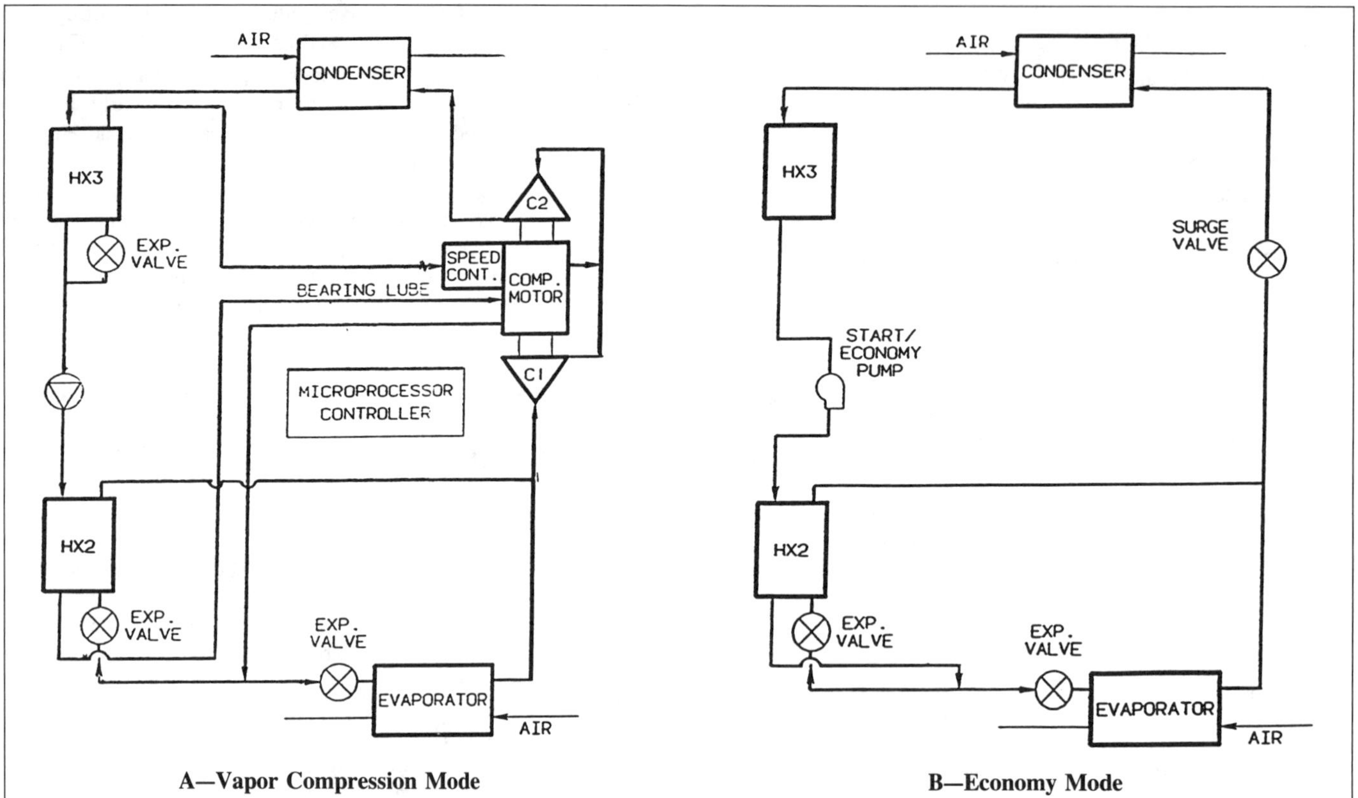

Fig. 4 Physical Implementation of Two-Stage Vapor Compression Cycle

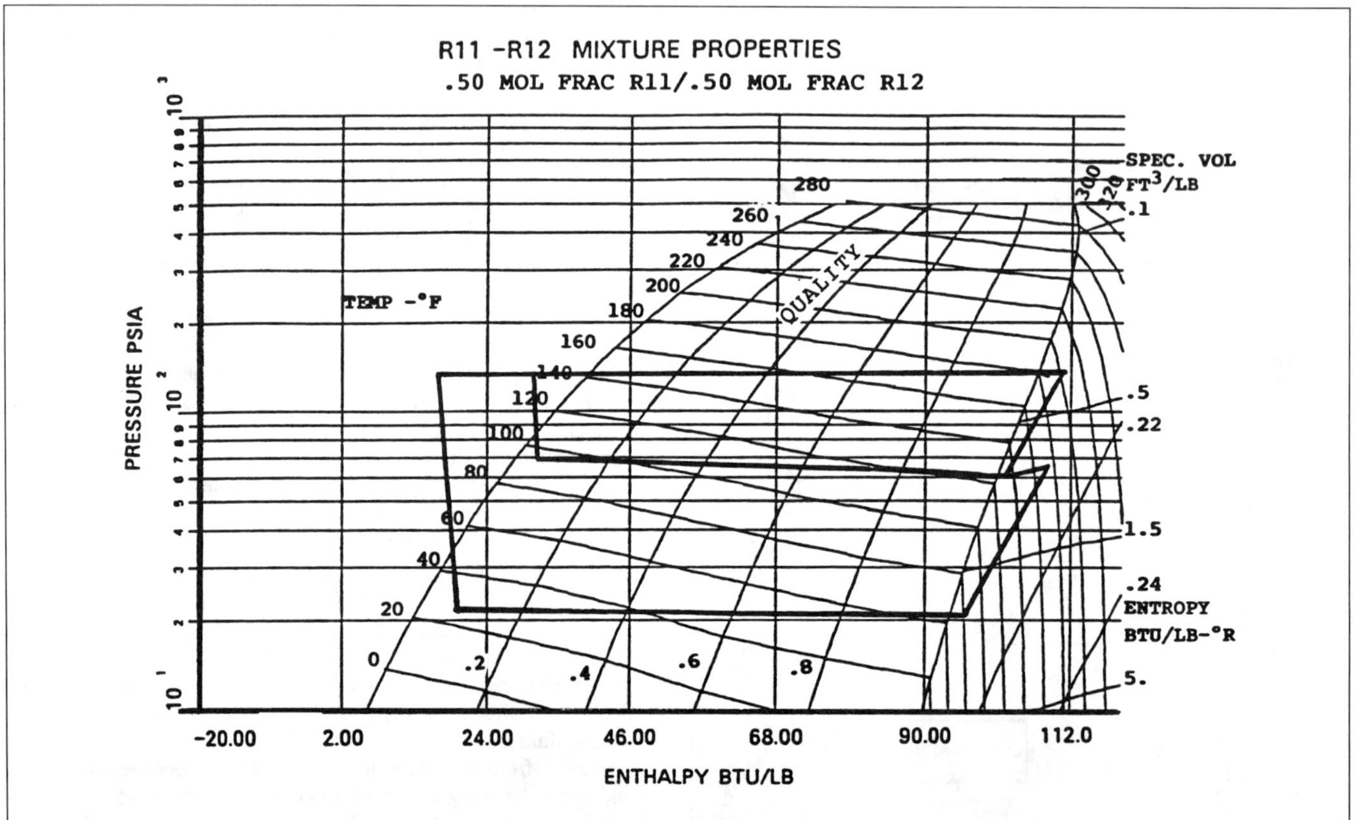

Fig. 5 State Points of NARM in Two-Stage Vapor Compression Mode

first-stage inlet while the evaporator expansion valve position controls delivered air temperature. Minimum superheat from HX2 is controlled by its expansion valve while minimal superheat requirement at the second-stage compressor inlet is controlled by varying the HX3 expansion valve position.

SYSTEM OPTIMIZATION

Having established an efficient cycle arrangement, design optimization required solution of refrigerant constituents, composition, and cycle states that would result in the "best" design. For this aircraft application, the goodness criterion was minimum take-off gross weight (TOGW). TOGW is the sum of hardware weight and the fuel weight associated with generating the compressor power (efficiency-dependent) and overcoming the drag resulting from using ambient cooling air for the condenser. For a static commercial or industrial application, the goodness criterion would most likely be defined in direct economic terms such as cost of ownership. Its components would include acquisition and operating costs with the same roots of size (weight) and efficiency. Selection of cycle states involves the classical balance between the desire for maximizing the COP obtained by maximizing the evaporator/condenser temperature ratio, and the resulting increasing heat exchanger sizes for fixed heat source and heat sink temperatures.

To accomplish the optimization process, a steady state computerized design system was developed. The logic flow is shown in Figure 6. The system includes models for the refrigerant, thermodynamic cycle, and sizing routines for all major system components. It allows the concurrent calculation of power draw, cooling flow, and system weight and resulting TOGW to be made for a given NARM and cycle conditions. The NARM mixture ratio and cycle conditions (evaporator and/or condenser temperature) may, in turn, be automatically sequenced to provide a complete matrix of possible system designs from which to choose the lowest TOGW solution. Each pass through the program requires component designs to be computed for the prevailing thermodynamic conditions. Thus, the compressor motor and impellers are sized for the power consumption, flow, and pressure ratio; the fans are sized for the cooling flow requirement of the condenser; and detailed heat exchanger sizing is performed based on internal configuration data. As shown in Figure 7, the nonlinear two-phase temperature gradient of NARM required a numerical approach for sizing the heat exchangers where the thermal conductance requirement was accumulated by stepping through the heat exchanger while solving the energy and heat transfer equations.

Several pure refrigerants and NARMs (binaries/compositions) were analyzed in the optimization program. Typical results of the optimization process giving best performance for aircraft systems are shown in Figure 8 where COP, weight, and ram airflow rate, which form the components of TOGW, are plotted against evaporator temperature and several refrigerants. These characteristics, based on a specified condenser temperature, show the locii of system designs where a 85 mole % R11/15 mole % R22 system operating at approximately -9°C (16°F) has the lowest TOGW of 1,330 kg (2,920 lb).

SYSTEM DEVELOPMENT

Separate, parallel development efforts were undertaken on system elements that were deemed high-risk. These included characterization and handling of NARM, the motor drive inverter and rectifier/filter, the compressor, and the microprocessor controller.

To obtain early experience, before committing to full-scale development, a large-capacity system was designed with commer-

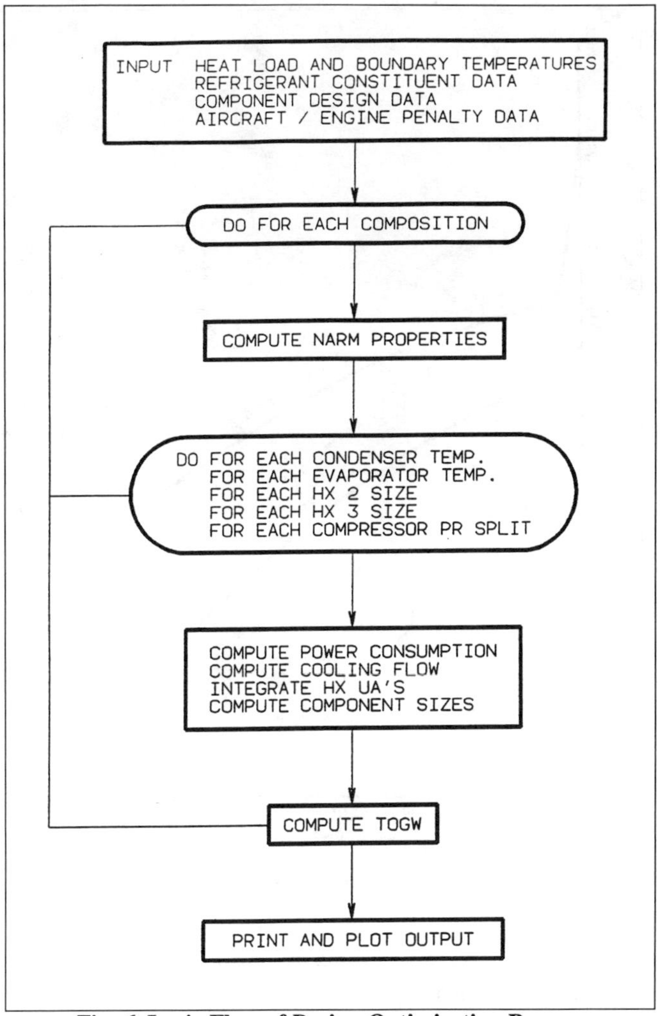

Fig. 6 Logic Flow of Design Optimization Process

cially available components to verify system performance, to measure temperature gradients in heat exchangers, and to develop techniques to mix and charge a system with NARM. Two positive-displacement industrial compressors separated by an electrical heating element simulated the two-stage, motor-driven compressor. Tube-in-tube heat exchangers represented the evaporator, condenser, and subcoolers. These were instrumented to measure pressure and temperature at intervals along their lengths. Of the various mixing and charging techniques examined, it was found that charging the system with the appropriate mass of the less volatile component followed by the more volatile component was simple and effective. The mixture came to equilibrium quickly in the system. The mixture ratio was subsequently verified by taking single-phase gas and liquid samples and subjecting them to gas chromatograph measurements where it was found that the mixture was stable and constant throughout.

The variable-speed brushless permanent magnetic motor that drives the compressor is controlled by a motor power supply (MPS). The MPS consists of a rectifier/filter unit and an inverter that uses bipolar junction transistors operated with PWM techniques to control the voltage and frequency applied to the motor windings. The MPS draws approximately 30 kW at full load and has to meet very strict limits on the amount of electromagnetic interference returned to the 400-Hz power source. Efficiencies at full load are 97 percent for the filter rectifier and 91 percent for the inverter.

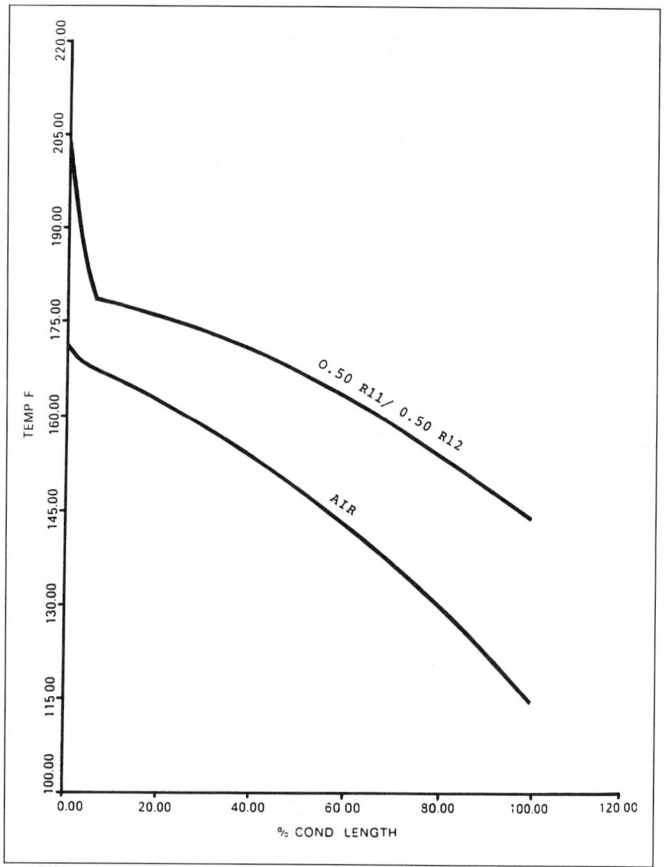

Fig. 7 Typical Temperature Profile in NARM Condenser

The combination of high speed (70,000 rpm) and high power (32 hp) of the 39-lb compressor assembly represented the highest power density motor in industry at the time of its design (1980) and presented a challenge in motor drive power electronic design. In addition, a requirement for very rapid motor acceleration was imposed by the existence of several critical speeds of the compressor shaft below 30,000 rpm. The drive is capable of accelerating the motor from 0 to 30,000 rpm in approximately 1 sec.

At the time design was started, the motor drive pushed the state of the art of power electronic technology, its cost was high, and a drive of this capability was not known to exist in industry. The past few years have seen very rapid advances in power electronic technology. Improvements such as the insulated gate bipolar transistor have reduced the cost of high-speed, high-power motor drives and have made them practical for a wider field of application. Future technologies such as metal oxide semiconductor controlled thyristors promise to reduce package size further and operate at even higher efficiencies.

The compressor shown in Figure 9 was developed on a dedicated rig designed to supply bearing lube and capable of setting a range of evaporator and condenser temperatures for calibration of the motor and each compressor stage. It was driven by a breadboard power supply developed during the electronics development program. Over 400 hr of successful operation was accumulated over the speed range of 30,000 to 70,000 rpm at loads of up to 25 kW before the design was committed to system testing.

Developing the microprocessor control algorithms was accomplished by generating a dynamic model of the system using the EASY5[1] (12) simulation package. This allowed various control strategies to be examined analytically. The EASY5 model was then used to represent the system in an AD10 high-speed computer real-time analysis where the gain and compensation of each control loop were determined to ensure stable operation at various rates of load application and changes in heat sink conditions.

These component development efforts culminated in 1987 with integrated system operation where additional development efforts were conducted to tune the controls for push-button operation and conduct preliminary performance analysis. Over 250 hours of system operation have been successfully demonstrated. Performance measurements were taken over the full ambient temperature range of -1 °C (30 °F) to 45 °C (114 °F) and full heat load capability of 42.2 kW (12 ton). Although a detailed performance calibration has yet to be performed, preliminary indications suggest that higher-than-anticipated bearing losses were evident and heat exchanger performance may have been adversely affected by maldistribution or orientation effects. A development plan was begun to explore and resolve these issues.

The system has been installed in a U.S. Navy P3 long-range patrol aircraft and is currently awaiting flight test.

ALTERNATIVE REFRIGERANTS

Apart from its inherent thermodynamic characteristics, consideration has to be given as to how a refrigerant affects the performance of the system components. Principal among these is the compressor. For systems that use centrifugal equipment, it is important to compress relatively high volumetric flow rates and to have adequately sized flow passages to minimize leakage and Reynolds number effects. For a given capacity (mass flow requirement), the desire for high-volume flow infers low-suction density conditions. However, it is also necessary to operate at an evaporator pressure greater than ambient to obviate inleakage of air, moisture, and other noncondensibles. This situation is particularly important in aircraft and shipboard applications with their relatively low driving speeds.

As an example, an existing analytic system design that used R114 to satisfy the high volumetric flow requirements of its centrifugal compressor while operating at superatmospheric conditions was reevaluated for use with the ozone safe replacements. Table 1 shows a comparison of system performance with R114 and new systems designs with the designated replacement R124 and a 65 mole % R123/35 mole % R134a NARM.

It is evident that, although the properties of R124 are closest to R114, R124 has a considerably higher vapor pressure, which has an adverse effect on compressor performance (considerably lower volumetric flow), resulting in a computed compressor power consumption that is 20 percent greater. R124 is also among the second tier of replacement materials being subjected to toxicity testing. To preserve the relatively high performance of the system using R114, an examination of designs using mixtures of tier 1 replacements was conducted. The obvious pair was R123 (R11 replacement) and R134a (R12). Using a mixture of 65 mole % R123 provided the required high volumetric flow at superatmospheric conditions and resulted in an efficient compressor design that restored system efficiency to above its R114 level. An added benefit of the NARM is clearly shown by reduced cooling airflow, which reduced total power draw from 39.2 kW to 37.2 kW, a 5-percent power savings.

[1] EASY5 is a registered trademark of Boeing Computer Services.

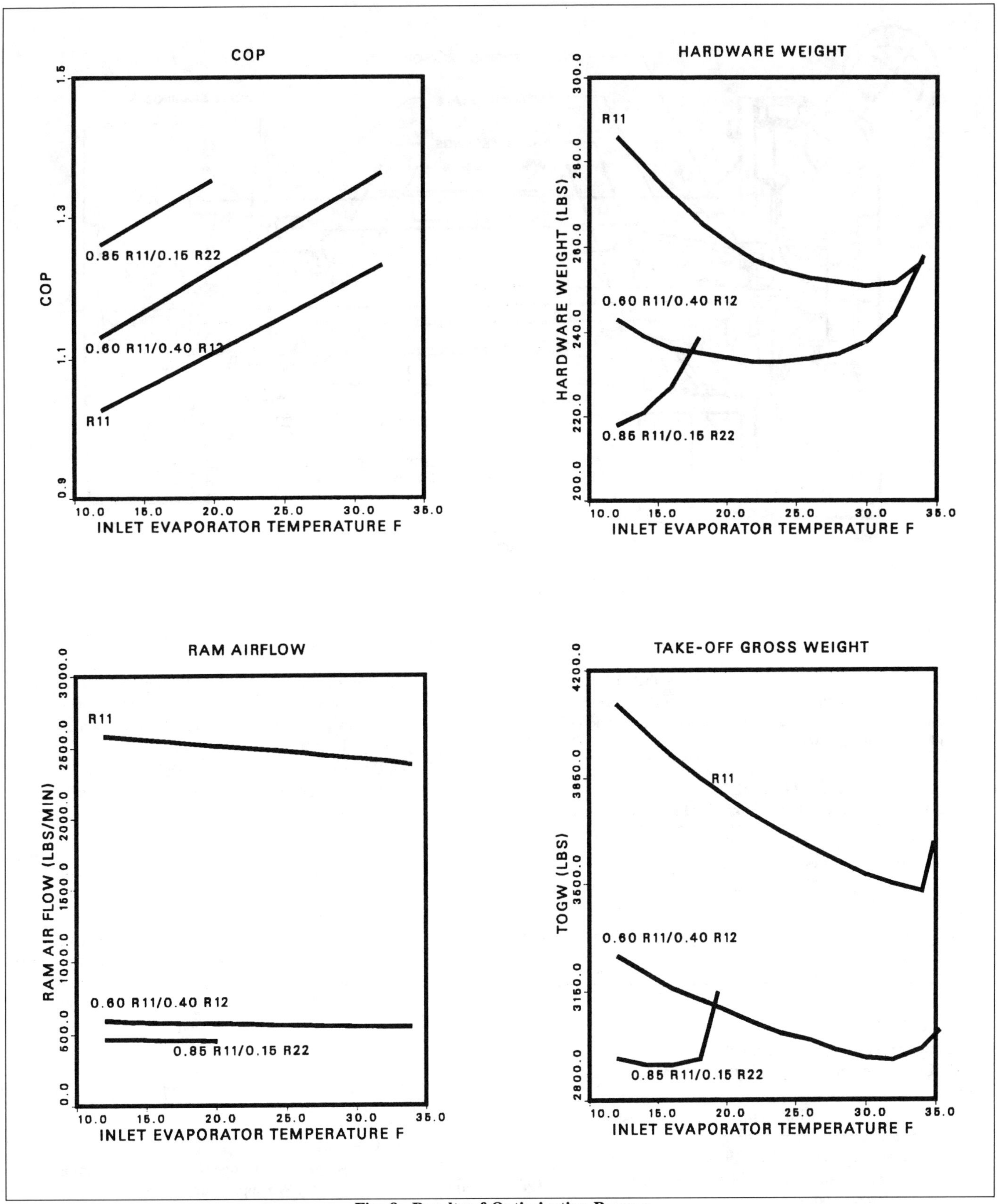

Fig. 8 Results of Optimization Process

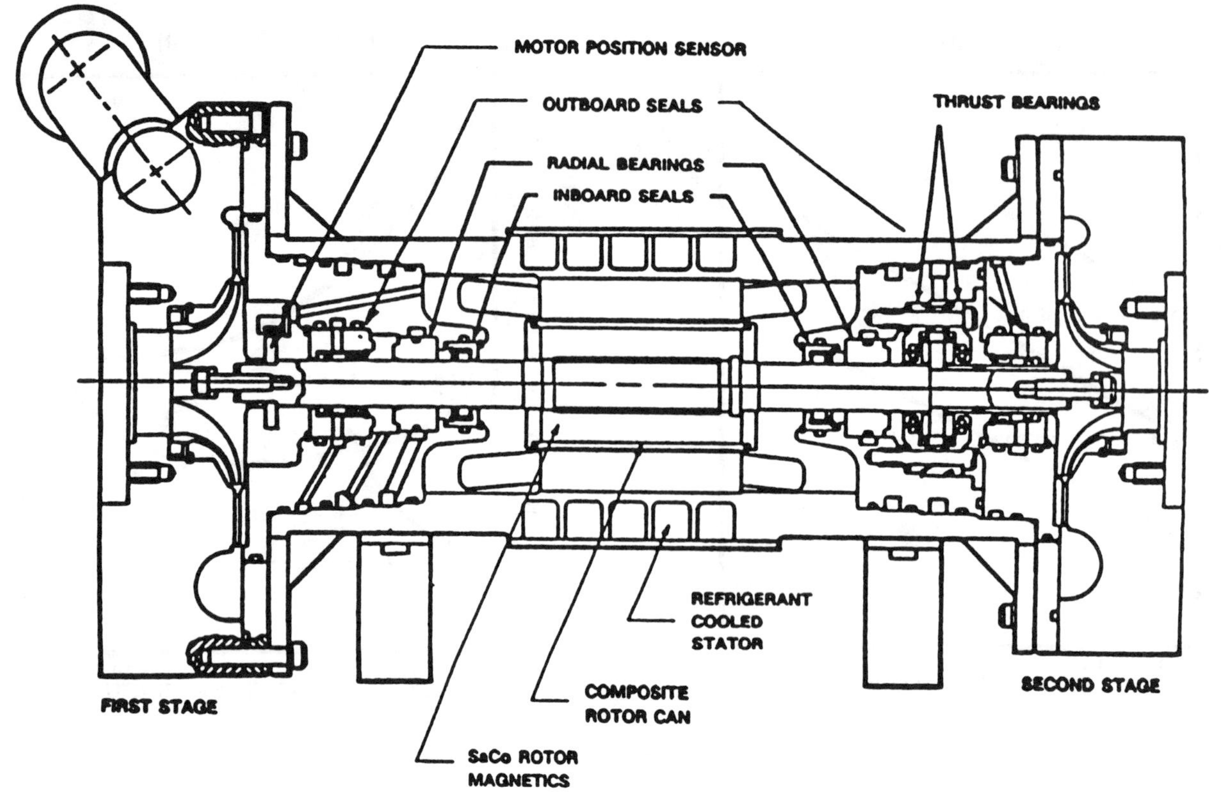

Fig. 9 Cross-Section of Refrigerant Compressor

This limited study clearly demonstrates that mixtures of new refrigerants can be applied with high efficiency to refrigeration applications and may offer the only alternative in replacing some of the to-be-banned substances.

Tab. 1 P-7A (42.2 kW/12 Ton)
Vapor-Cycle Cooling System Performance Analysis

Refrigerant	Evaporator Pressure (psia)	Condenser Airflow Rate (lb/min)	Power Draw		
			Comp (kW)	Comp Fan (kW)	Total (kW)
R-114	15.0	321	32.5	6.7	39.8
HCFC-124	24.3	382	39.5	8.0	47.5
65% HCFC 123/ 35% HFC 134a	15.1	262	31.7	5.5	37.2

CONCLUSIONS

Our experience in the design, optimization, and development of a NARM vapor compression cooling system has lead to the following conclusions:

1. Using the NARM thermodynamics property and system cycle analysis computer programs we developed in the early 1980s, we successfully designed and predicted the performance of a vapor-cycle cooling system using a nonazeotropic refrigerant mixture.
2. We have successfully operated a vapor-cycle cooling system using a NARM with no significant chemical, mechanical, or mixture problems and have met predicted performance.
3. Indications are that an advanced design NARM vapor-cycle cooling system can save considerable energy over presently used air-cycle and vapor-cycle systems using a single component refrigerant.
4. NARM offers considerable flexibility in the design and operation of a system. For example, using a suitable NARM, it is possible to develop a cooling/heating system operating at a low-pressure ratio and giving good performance. The low-pressure ratio allows using a single-stage compressor with a consequent increase in reliability and simplicity. The reader may refer to (13) for a description of this system.
5. With the demand from increased cooling loads on modern aircraft, along with the demand for fuel-efficient, no-bleed engines, vapor-cycle cooling systems will receive increased consideration as a replacement for air-cycle systems.
6. Economy mode operation available with this type of cycle (using only a liquid recirculation pump) will provide considerable energy savings.
7. Further improvement in system performance is possible by using magnetic bearings in the compressor. Sundstrand is pursuing this technology.

ACKNOWLEDGEMENT

The authors wish to acknowledge the many contributions of their colleagues at Sundstrand. We appreciate and thank Messrs. Anson S. Coolidge and Jay Fernandes for their active support of this program from the beginning. The interest of Mr. James E. McNamara, U. S. Naval Air Development Center, and Mr. George Stanton, Lockheed Aeronautical Services Corporation, is also gratefully acknowledged.

REFERENCES

1. Troxel, S. and Braven, K. D., "The Use of Nonazeotropic Refrigerant Mixtures in Residential Heat Pumps," Advances in Heat Pumps, Presented at the Winter Annual Meeting of the ASME, San Francisco, California, pp. 9-14, December, 1989.
2. Cloud, W. W., McNamara, J. E., and Wigmore, D. B., "A New Technology in Energy-Efficient Electrically Driven Aircraft Environmental Control Systems," 21st IECEC Conference, August, 1986.
3. McLinden, M. O. and Radermacher, R., "Methods for Comparing the Performance of Pure and Mixed Refrigerants in the Vapor Compression Cycle," Int. J. Refrig., Vol. 10, pp. 318-325, November, 1987.
4. Radermacher, R. and Lavelle, J., "Comparison of Refrigerant 12 With the Blend of R-22/R142b," Proceedings of ASME Winter Annual Meeting, Chicago, Illinois, pp. 25-30, November, 1988.
5. Domanski, P. A. and McLinden, M. O., "A Simplified Cycle Simulation Model for the Performance Rating of Refrigerants and Refrigerant Mixtures," ASHRAE, Purdue Refrigeration Conference, West Lafayette, Purdue University, pp. 466-475, 1990.
6. Kruse, H. H., "Current Status and Future Potential of Nonazeotropic Mixed Refrigerants," Proceeding of the 1987 International Energy Agency Heat Pump Conference Prospects in Heat Pump Technology and Marketing, Orlando, Florida, April, 1987.
7. Mokadam, R. G., "High Efficiency Refrigeration or Cooling System," U. S. Patent No. 4,598,556, July 8, 1986.
8. Soave, G., "Rigorous and Simplified Procedures for Determining the Pure-Component Parameters in the Redlich-Kwong-Soave Equation of State," Chemical Engineering Science; Vol. 35, N8, pp. 1725-1729, 1980.
9. Soave, G., "Equilibrium Constants From a Modified Redlich-Kwong Equation of State," Chemical Engineering Science, Vol. 27, pp. 1197-1203, June, 1972.
10. Reid, C. R., Prausnitz, J. M., and Sherwood, T. K., The Properties of Gases and Liquids, McGraw-Hill, Inc., 1977.
11. Morrison, G. and McLinden, M. O., "Application of a Hard Sphere Equation of State to Refrigerants and Refrigerant Mixtures," NBS Technical Note 1226, National Bureau of Standards (National Institute of Standards and Technology), August, 1986.
12. "Engineering Analysis System (5th Generation)" made by Boeing Computer Services.
13. Mokadam, R. G., "Low Pressure Ratio High Efficiency Cooling System," U. S. Patent No. 4,809,521, March 7, 1989.

CONCEPTUAL STUDIES ON MODULAR ADSORPTION SYSTEMS FOR VARIOUS APPLICATIONS

K. F. Knocke, A. Gerlich, and R. Strauss
Institute for Technical Thermodynamics
Aachen, Germany

ABSTRACT

At the Institute for Technical Thermodynamics, at the University of Aachen, a new sorption machine is under investigation that consists of many periodically operating adsorption units. The planed system zeolite/water enables high temperature lifts. For cooling applications the system activated-carbon/methanol is interesting.

For the layout of the adsorption machine a simulation program has been developed. Especially the COP obtained in different machine concepts concerning the internal heat exchange has been studied under various boundary conditions.

The presented machine design is similar to rotary regenerators. All sorption units are arranged on a slowly moving rotor. Heat transfer agents are flowing continuously through the rotor. The COP can be significantly improved by internal heat exchange. The machine is suitable for heat transformer, heat pump and air conditioning applications and can be build in a very simple way without solution pump and expansion valves.

INTRODUCTION

In the field of energy saving the possibilities of sorption machines are not used sufficiently. One reason is the difficulty to build machines which are favourable from the thermodynamical and the economical point of view. Most of the thermodynamically sophisticated systems presented in the past are too expensive to achieve interesting pay back times. Especially in systems with a power of less than 20 kW expensive components such as electric solution pumps or complicate control systems should not be used. For this application modular periodically working machines seem to be a good solution. The lower coefficient of performance (COP) of a single periodically working unit compared with a continuously working system can be improved by the interconnection of several units ormodules. By interconnecting different numbers of modules various systems can be installed with the same components. Therefore it is not necessary to develope one special machine for each application and power range. This may lead to an economical system for space heating and space cooling.

WORKING PRINCIPLE

The sorption modules consist of two connected vessels. One vessel contains the sorbent and operates as adsorber and desorber. The other one is the evaporator and condenser for the sorbate. The working principle for heating application is shown in Figure 1.

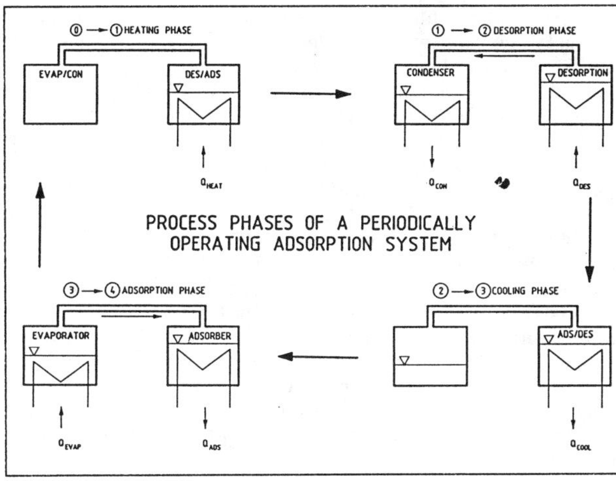

Figure 1: Working principle of the periodical sorption process

At the beginning of the cycle the condenser/evaporator is empty. By the heat input Q_{Heat} the ad-/desorber is heated up until the process pressure reaches the condensation point in the condenser. Then refrigerant is desorbed by input of the heat of desorption Q_{Des}. The refrigerant condenses in the condenser emitting the useful heat of condensation Q_{Con}. When a certain temperature or concentration in the desorber is achieved, heat input in the desorber is switched off. During the next phase the ad-/desorber is cooled down delivering the heat Q_{Cool}. When the temperature in the evaporator reaches the temperature of the ambient heat source, the sorbate is evaporated consuming low temperature heat Q_{Evap}. Produced sorbate steam is adsorbed by the sorbent while the heat

Q_{Ads} being produced is transferred to a heat transfer medium. The cycle can start again when all the refrigerant is expelled from the evaporator (Knoche, Seitz, Zerres, 1991).

SORPTION SYSTEMS

For this type of machine any sorption system with high difference in boiling points between sorbent and sorbate can be used. It has to be ensured by a good construction that no sorbate gets into the condenser/evaporator.

For most applications zeolite/water systems have a lot of advantages. The sorption properties are very good. The useful temperature range is large at excellent thermal stability. Since water has a very high enthalpy of evaporation the specific heat input per kg sorbate is very high, too. Due to the low process pressures the walls of the vessels can be thin and light. This is quite important for periodically working systems. Heavy vessels require a lot of heat for heating up, reducing the COP. This is an important disadvantage of for example ammonia systems for this type of machine. The type of zeolite that should be utilized depends on the application. For heating and cooling the best results can be achieved with the X-type zeolites like the 13X. At high temperatures it is better to use A-type zeolites like the CaA or MgA due to their higher thermal stability (Rothmeyer, 1985). High temperatures are interesting for heat-transforming applications or high-temperature heat pumps. But also in the designed heat pumps it is possible that temperatures of more than 200°C may be achieved at the walls in the desorption phase. The calculations of different machine concepts described in the next chapter with a MgA zeolite are based on data measured by Maier-Laxhuber (1983).

Another interesting sorption system is the working pair activated-carbon/methanol. The measured sorption properties (Bock et al., 1987) seem to be much better than the theoretically calculated (Pohl, 1986). Already the calculated properties result in a good COP. A main problem is the thermal instability of methanol. Literature reports quite different temperatures of decomposition (Pons, Guilleminot, 1986). Our measurements showed instability of methanol in contact with activated-carbon already at a temperature of 130°C. Activated-carbon/methanol may be a good working pair for cooling application but it is not suitable for the utilization in heat pumps or heat transformers.

DIFFERENT CONCEPTS OF MODULAR HEAT PUMPS

The desired heat pump will consist of many simple adsorption modules; each one working with a low performance. By internal heat exchange from one module to another the heat input at high temperature can be reduced. With this the output of the heat of adsorption decreases, respectively. But due to the constant heat output from the condenser (heat pump application) or heat input in the evaporater (cooling application) the COP increases.

Internal heat exchange can be realized in many different systems. Trying to achieve the highest COP leads to a machine that will never be acceptable from the economical point of view. Thus it is necessary to design and compare different more or less complicated and expensive concepts. For heat pump application six different concepts will be discussed. The modules themselves consist of two vessels as described above.

For the calculation of the different processes a steady state simulation program has been developed. It includes routines for the calculation of the thermodynamical parameters of different sorption systems. The simulation program enables to investigate the behaviour of each single module and of the combination of all units. Next the numerical models will be used in an dynamic simulation program that has been developed at the Institute for Technical Thermodynamics for absorption heat pumps (Knoche, Zerres, Seitz, 1990).

The following comparison of different heat pump concepts bases on a machine that consits of n modules. After the cycle-time t_{cycl} every module is back to the same state as at the beginning of the process. In a certain time-period $dt=t_{cycl}/n$ every module passes one process step. Every process step is determined by the properties of a module at the beginning and at the end of this step. When the module with the number k reaches the end of a process step it is right at the beginning of the next one. Module k becomes module k+1 and on the other hand module k-1 becomes module k. So after each process step the place of the modules in the process has been changed but the state of the whole machine is the same again.

The temperature in the adsorber at the end of the adsorption phase and in the condenser is the useful temperature T_{use}. The pressure drop between evaporater and adsorber due to the mass transfer has been neglected. The transfered heat should be nearly the same in any process step. Thus the change of concentration in the adsorption or the desorption steps is constant. In the cooling down or heating up phase the temperature steps are constant. For the heat transfer a minimum temperature difference can be fixed, depending on the construction.

The simplest machine, concept I, that can be build consists of modules without any connection one to another. The modules should work in different phases at the same time in order to get nearly stationary heat fluxes coming in and out of that machine as shown in Figure 2. The heat of adsorption Q_{Ads} and of cooling down Q_{Cool} can be used together with the heat of condensation Q_{Con} for the heating application. The burner supplies the heat of desorption Q_{Des} and for heating up Q_{Heat} after the adsorption phase.

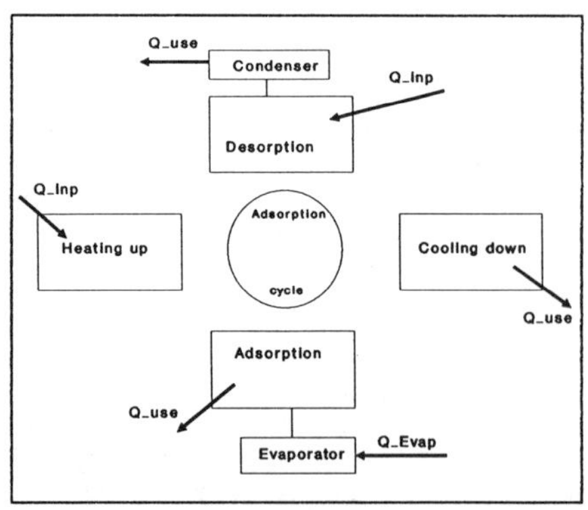

Fig. 2: Machine concept without internal heat exchange

Such a system is quite cheap and has a low COP. The COP_1 can be calculated as:

$$COP_1 = \frac{Q_{use}}{Q_{inp}} = \frac{Q_{Con}+Q_{Ads}+Q_{Cool}}{Q_{Des}+Q_{Heat}}$$

The first step to improve the COP is to exchange heat for cooling down and for heating up (Concept II). Since the temperature levels are quite different it can be realized easily. For this heat exchange at least two modules or a regenerative heat exchanger are necessary. In order to avoid "dead times" concerning the external heat transfer two more modules are needed. With this first internal heat exchange an improvement of the COP of about 5-15% can be achieved. As both heats, Q_{Cool} and Q_{Heat}, are of the same order of magnitude an estimation of COP_2 is possible:

$$COP_2 \approx \frac{Q_{Con}+Q_{Ads}}{Q_{Des}} = \frac{Q_{Con}}{Q_{Des}} + \frac{Q_{Ads}}{Q_{Des}} \approx \frac{Q_{Con}}{Q_{Des}} + 1$$

In order to get a further improvement of the COP the heat input for the desorption process has to be reduced. First this can be realized using heat of cooling down for the desorption. Then the heat for heating up the modules is supplied from the adsorption phase. Second, according to the marginal conditions a part of the heat of adsorption can be utilized for the desorption, as well.

This can be seen clearly in the schematic vapour pressure diagram, figure 3. In the winter season the temperature of evaporation $T_{e,w}$ is low and the usefull temperature $T_{u,w}$ is high. At this time the desorption temperature is always above the adsorption temperature. Thus no heat of adsorption can be transfered to the desorption phase. But most of the heat of cooling down can be utilized for the desorption. In spring time or in autumn the evaporation takes part at a higher temperature level $T_{e,s}$ and the usefull temperature $T_{u,s}$ is lower than in the winter. The large overlap in the adsorption and desorption temperatures enables the transfer of considerable parts of the heat of adsorption into the desorption phase.

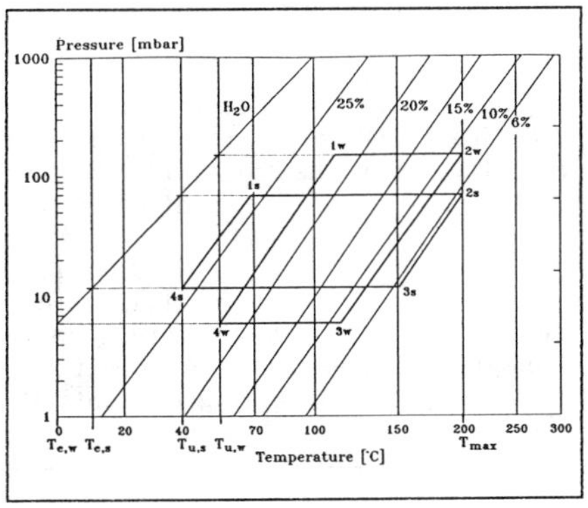

Fig. 3: Sorption cycles in a pressure diagram for zeolite(Mga)-water; winter: w, spring: s

The maximum theoretical COP_3 is more than 50 % better than COP_2. The realization of such a machine-concept, concept III, is extremely complex. Figure 4 gives an idea of the necessary connections between a large number of modules. The control of all these fluxes requires a complicated system. Thus such a concept is not realistic but shows the upper limits in terms of energy conversion.

In the following concepts all heat coming in and out of the vessels is transferred by one or two heat transfer medium fluxes to achieve a COP comparable with COP_3 in a realistic manner.

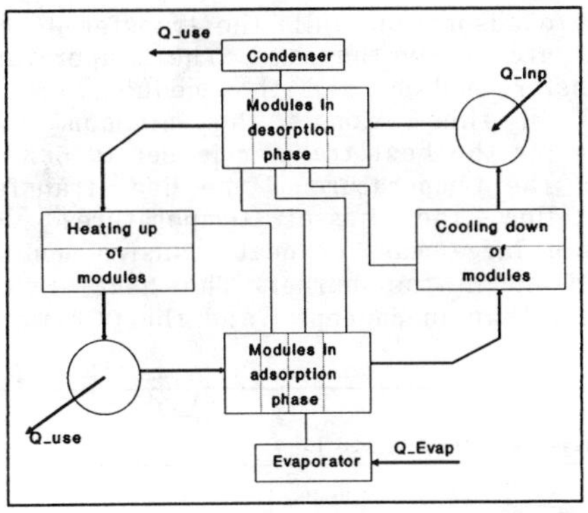

Fig. 4: Machine concept with ideal internal heat exchange

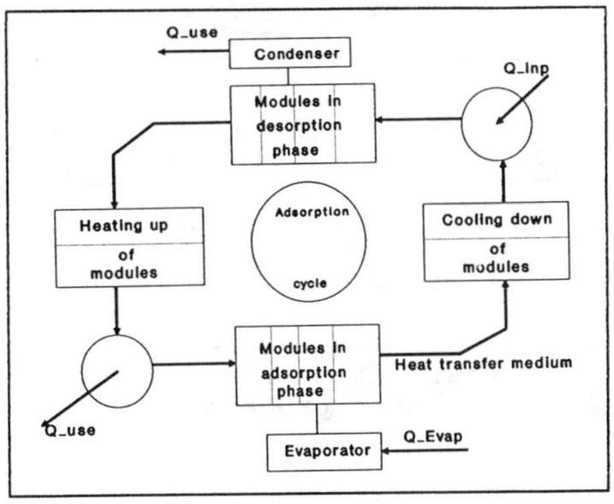

Fig. 5: Machine concept with one heat transfer medium circuit

In concept IV one heat transfer medium circuit is designed as shown in Figure 5. The medium flows in the opposite direction to the process and passes all modules, through the burner and the heat exchanger of the heating system. It leaves the heat exchanger with nearly runback temperature and passes the modules that are in the adsorption phase, beginning with the coldest and ending with the one that just started adsorption. Then it passes the modules that cool down. On this way the medium is heated up without reaching the temperature required for the desorption. It has to be heated up additionally by a burner, for instance. Then the medium performes the desorption and the heating up of the last vessels. Finally it has to be cooled down in the heat exchanger giving useful heat to the heating system. This heat is of the same order of magnitude as the heat brought into the process by the burner at the high temperature level. The heat coming in or out of the different phases of the process is not proportional to the process temperature. Due to this different fluxes of heat transfer medium are required in every process step. To enable the heat transfer in all modules the temperature difference between the module and the heat transfer medium should not fall below the fixed minimum. Therefor the mass flux of heat transfer medium has to be adequately high. The dependency of the required heat transfer medium for a 26 modules heat pump is shown in Figure 6.

Here the substantial differences in the adsorption and desorption phase are demonstrated. This results from the steeper temperature slope during desorption compared to adsorption while the transfered heats, Q_{Ads} and Q_{Des}, are nearly the same. The temperatures of the heat transfer medium and the modules are shown in figure 7. In this example the minimum temperature difference for the heat transfer is set to 5 K. It can be seen that the temperature of the heat transfer medium cannot follow the vessel temperatures very well. Therefore a large mass of heat transfer medium has to be heated up in the burner. The heat input may be even larger than in concept I and the COP worse.

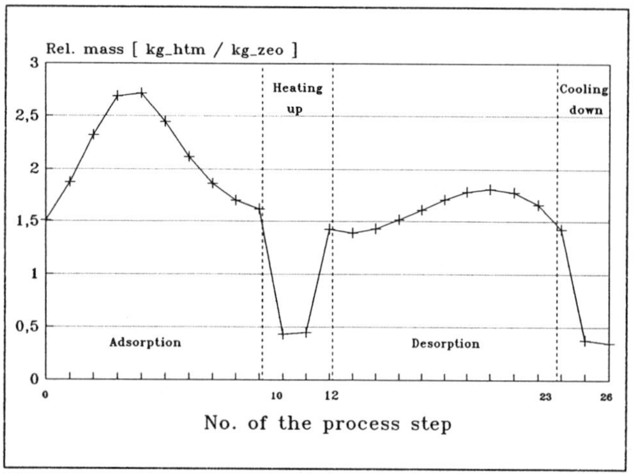

Fig. 6: Nessecary heat transfer medium during the process

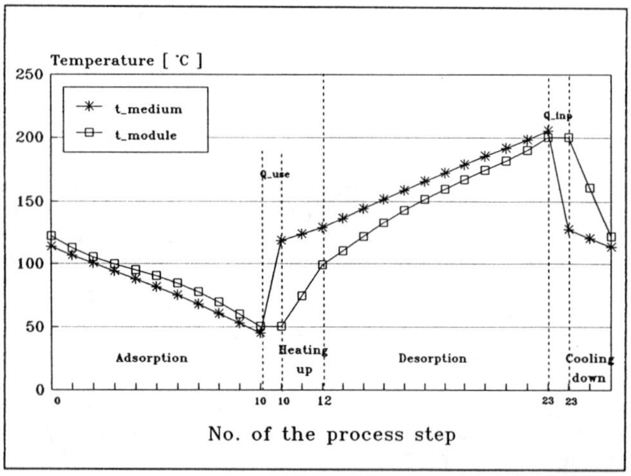

Fig. 7: Temperatures of modules and heat transfer medium during the sorption cycle

In order to avoid this disadvantage splited circuits can be used. Only the mass needed for desorption is heated up. This can be realized by two separate circuits (concept V), one for the adsorption and one for the desorption as demonstrated in Figure 8. Another possibility is the separation of the mass flux after the adsorption phase and recombination after the desorption phase (concept VI) as shown in Figure 9. Both concepts require a simple control system and concept V a second pump for the heat transfer medium.

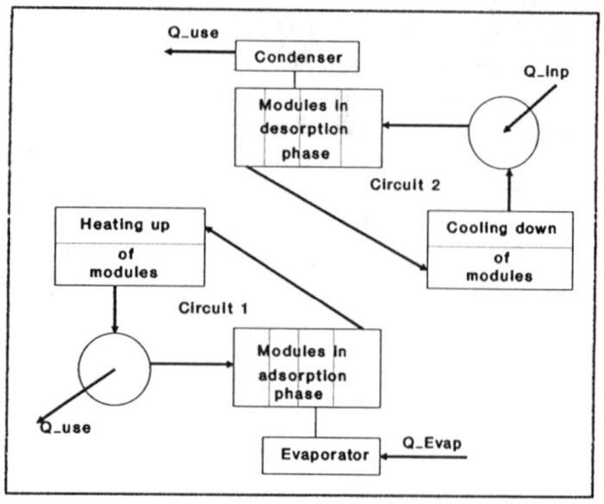

Fig. 8: Machine concept with two circuits of heat transfer medium

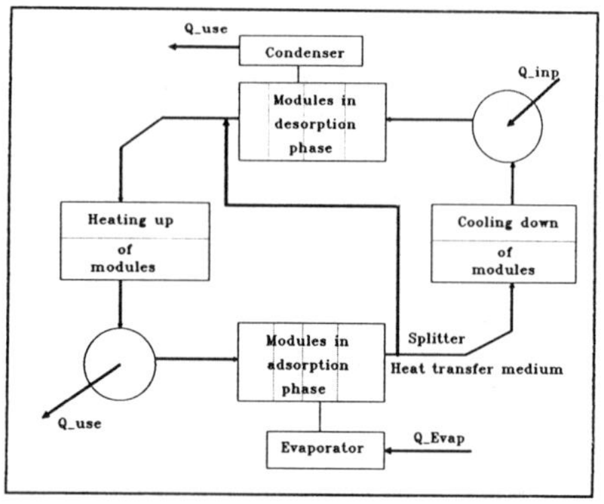

Fig. 9: Machine concept with splitted heat transfer medium

For space heating application the heat pump has to work with high COP over a wide range of ambient temperatures T_{ev}. The useful temperature T_{use} also depends on the design of the heating system. For the different concepts some calculated COP are shown in the Figures 10 to 12. The temperature difference for heat transfer dT is 5 or 10 K. The maximum process temperature is set to 200°C. The heat storage capacity of the vessels is neglected because the effect is not very important for the more interesting concepts III to VI.

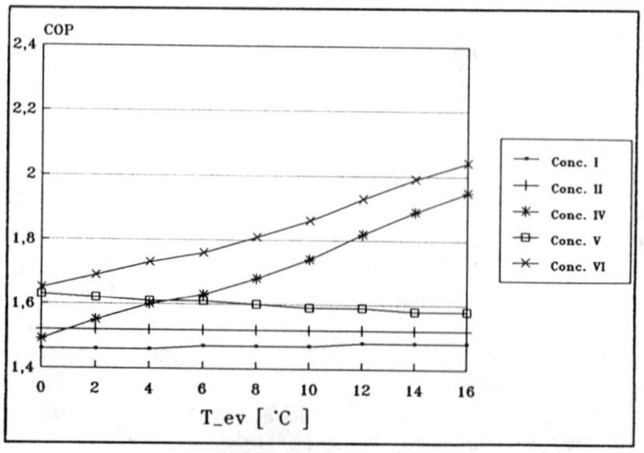

Fig. 10: Calculated COP of different machine concepts; T_{use} = 50°C, dT = 5 K

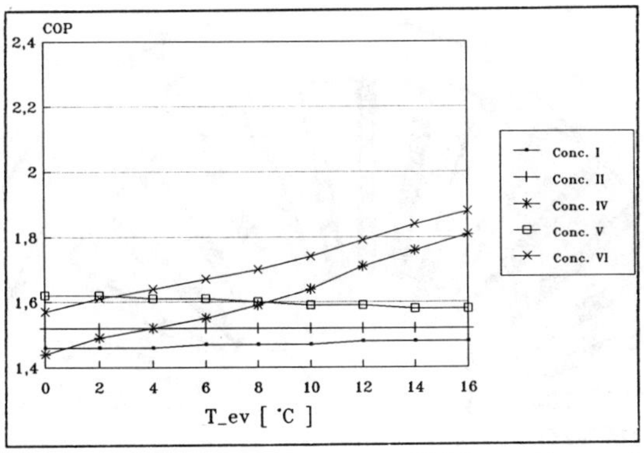

Fig. 11: Calculated COP of different machine concepts; T_{use} = 50°C, dT = 10 K

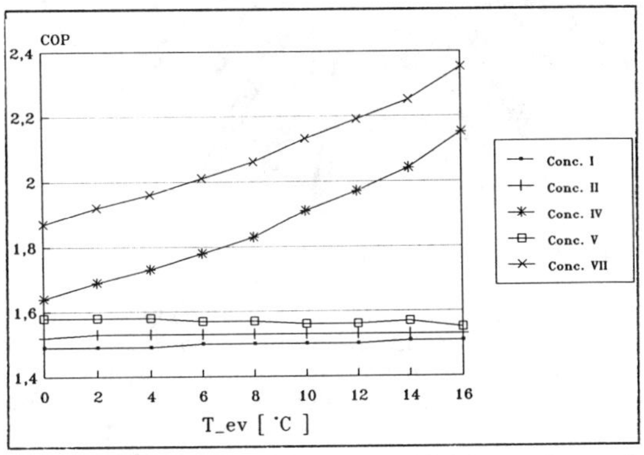

Fig. 12: Calculated COP of different machine concepts; T_{use} = 35°C, dT = 10 K

It can be seen that COP_1 and COP_2 are nearly constant in the overall temperature range. The COP of the two-cycle-system COP_5 is only about 10 % better than COP_2 and decreases about 10 % with decreasing temperature of condensation as well as with increasing ambient temperature. This results from the increasing temperature difference between the beginning of adsorption and beginning of desorption. In the other concepts this leads to a big increase of the COP due to the heat transferred from adsorption to desorption. For high useful temperatures, low ambient temperatures, and under consideration of the necessary temperature difference for the heat transfer concept V is even better than concept IV.

Resuming the results concept VI can be seen favourable for a modular adsorption heat pump for space heating application with a zeolite MgA-water system. Only in the winter season and for necessary high temperature differences for heat transfer depending on the construction concept V is interesting. In this case a combination of systems IV and V seems to be promising. This combination can be realized by two standard valves which combine or divide the two circuits. In winter the system would work with two circuits; in the rest of the year only with one. The switch over from system IV to system V can be controlled by the temperatures of the heat transfer medium after the adsorption phase and after the desorption phase.

DESIGN OF AN ADSORPTION ROTOR

Based on the thermodynamic studies of the first part the construction design of an adsorption machine will be presented now. A special feature is the fact that it is built up out of many small, periodically operating sorption units.

Build-up of a single sorption unit

The basic construction of a single sorption unit is shown in figure 13. The unit mainly consists of an evaporator/condenser and an adsorber/desorber section. In the upper section the refrigerant (water) can be stored inside the capillary meshes. These meshes are attached to the heat exchanger walls. Zeolite extrudates, which are located inside the adsorber/desorber section, are fixed by a perforated plate. The heat fluxes converted in these two sections are transmitted by heat exchangers to the fluids A and B. Thermal conduction between evaporator-/condenser and adsorber/desorber section is avoid by an insulation.

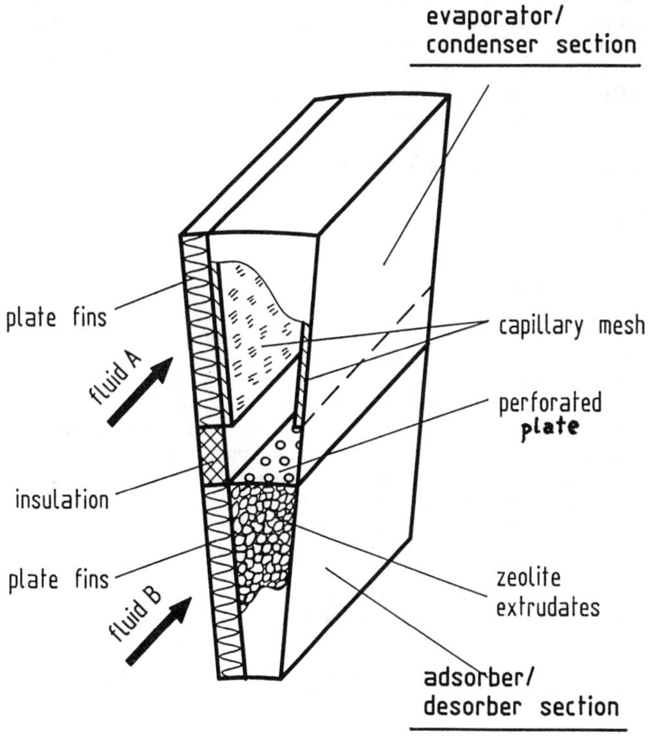

Figure 13: Basic Construction of a Sorption Unit

The units operate according to the closed, periodical sorption cycle explained in chapter 2. Each cycle is composed of alternating adsorption and desorption phases. The temperature levels of the two heat transfer agents determine the actual phase of a unit. Due to this the working pair is meant to even out the imbalance (which is forced by the temperature levels) by changing the concentration of the refrigerant. So, the sorption unit itself is passive with regard to the state of process, it cannot control the process.

The construction of an evaporator/condenser and adsorber/desorber section is the simplest way of realising a periodic adsorption machine. No solution pump, expansion valves and connection pipes are necessary. As a result the single modules can be sealed easily.

From the constructive point of view the units mainly exist of heat transfer areas. As heat exchanger - beside a pipe design - brased plate heat exchangers as shown in figure 13 are suitable. The heat transfer agents flow through two parallel thin plates, which are brased with corrugated plate fins. This kind of heat exchanger has a very low weight related to its capacity. This is very important for a periodically operating adsorption machine, because high thermal storage capacities influences the COP negatively. Next to the low weight these heat exchangers are very compact (they are often called `compact heat exchangers`) and have remarkably heat transfer coefficients. Furthermore all kinds of fluids can be used and the dimensions of the units are variable to a large extent.

In order to attain reasonable heat transfer coefficients the heat transmission between zeolite and heat transfer area is of deciding importance. In this context we are building up test rigs to examine the heat conductivity and the kinetics of adsorption on zeolites. Different types of binders and zeolite/wall connections will be tested. The aime of these examinations above all is to find out a zeolite structure with a sufficiently high heat transfer coefficient.

The adsorption rotor

Next to a lightweighted construction design the internal heat exchange is important for a good COP of a periodically operating adsorption machine. For continuously operating sorption machines this internal heat exchange can be realised easily e. g. by a solution heat exchanger or absorber generator heat exchange. An internal heat recovery is only possible for periodic adsorption machines, when inside the machine different sorption phases exist at the same time. This can be realised by a `thermal wave` (Tchernev and Clinch, 1989) or by many small sorption units (Niebergall). If sorption units are used, the connection of the fluid side is of importance.

A new development (Kaubek and Maier-Laxhuber, 1983), which is characterized by a simple fluid connection of the single units, is shown in figure 14. Many of the sorption units (see figure 13) are connected to a rotor, whose outer shape looks similar to rotary regenerators.

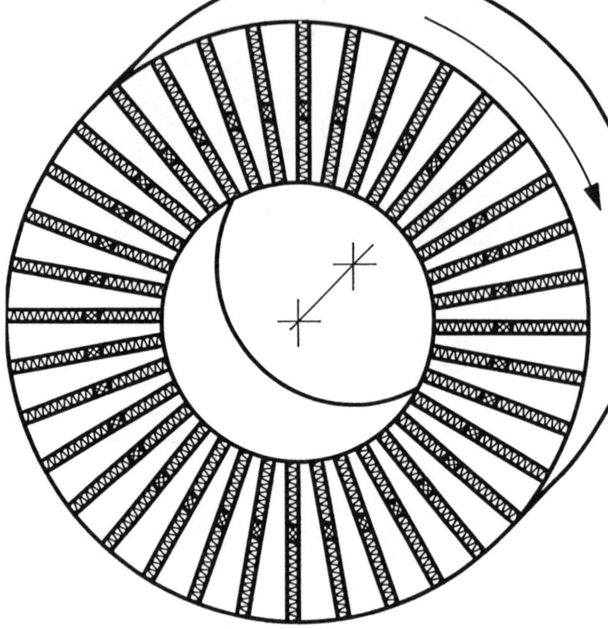

Figure 14: Principle design of an Adsorption Rotor

This adsorption rotor turns slowly round while heat transfer agents on different temperature levels - according to the different processes evaporation, adsorption, condensation and desorption - flow continuously through the heat exchangers. In figure 15 a possible fluid distribution of the inlet area is shown. This distribution corresponds to the simplest way of operation of a sorption machine without internal heat transfer.

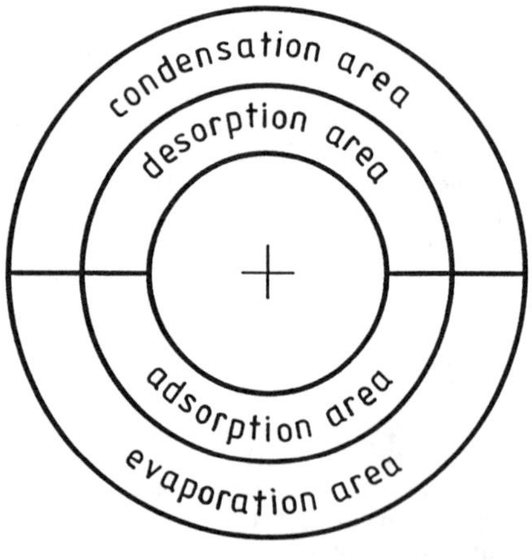

Figure 15: Fluid distribution of the inlet areas

The inlet and outlet areas do not rotate. In figure 16 is shown the principle design of an adsorption rotor with inlet and outlet ducts. Besides the different temperature levels of the fluids represent a heat pump application are shown. The heat transfer agents pass the heat exchangers in the single pass crossflow.

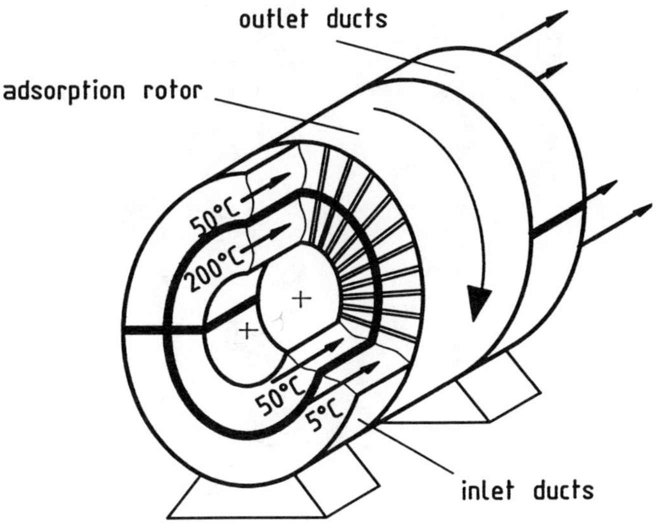

Figure 16: Principle design of an adsorption rotor system

To get a better idea of the working principle of the rotor, one sorption unit will be followed on its way during one rotation. A unit located in the upper half of the rotor is in the desorption phase, according to the temperatur levels of the fluids. While this unit passes the upper half, refrigerant steam is generated from the zeolite and liquefied in the condenser. When the sorption unit moves to the lower half, the desorption phase stops and the unit changes to the adsorption phase. In the beginning the unit cools down to adsorption temperature. Then the adsorption starts by evaporation of condensate inside the evaporator. The refrigerant concentration of the zeolite increases until the unit changes back to the upper half of the adsorption rotor. After heating up to desorption temperature, desorption of zeolite starts. After one rotation the unit is again in the same state as above mentioned.

A great advantage of the rotary principle are continuous heat fluxes by the heat transfer agents (assuming a sufficient number of units), although each unit operates periodically. So, a special feature of adsorption rotor systems are continuous capacities without solution pump and expansion valves, necessary for continuous absorption machines.

The internal heat exchange

The COP of adsorption rotors can be improved in the same way as well known from continuous absorption machines. All concepts studied in chapter 4 can be realised with adsorption rotor systems. Instead of the single pass crossflow of the fluids, as shown in figure 16, a multipass crossflow (approximated counterflow) is required. The optimal concept for the internal heat exchange can be found theoretically by following the second law of thermodynamics. But this solution is not necessarily the cheapest, see chapter 4. The rotary concept enables to test many different concepts of internal heat exchange easily by varying the inlet and outlet ducts. The future work will be an investigation in terms of the best concept from the economical point of view.

Control

If sorption machines are meant to operate at different temperature levels (operating points), it is important for the COP of the machine that the sorption process can be adapted to the different operating points. Single modules of an adsorption rotor are able to adjust automatically. This is achieved by the module design of two sections (evaporator/condenser and adsorber/desorber section) and by the ability to store refrigerant in the capillary meshes inside the evaporator/condenser section. So, adsorption rotor systems do not require costly components such as controllable solution pump and expansion valves, which are necessary for continuous absorption machines.

The capacity of rotor systems can be controlled by the rate of zeolite mass flow. This flow rate is easily controllable by the number of revolutions per time. As heating equipment an on/off controlled burner can be used, even for variable capacities. Furthermore adsorption rotor systems are well suited for direct air evaporators, because frozen heat exchanger areas thaw automatically when operating as condenser. No defrosting control is necessary.

Apart from the adsorption mode, adsorption rotor systems can be operated for domestic heating even on cold winter days. At low ambient temperatures, when the required temperature lift between evaporator and adsorber exceeds the possible temperature lift of the working pair (frozen refrigerant), only an insufficient adsorption process is possible. In that case adsorption rotors operate in the `direct heating mode`. The main heat transfer while operating in a direct heating mode takes place between desorber and adsorber. Because of the high temperature lift no heat transmission is possible between evaporator and adsorber. The change over from adsorption to direct heating mode takes place in a very simple manner by increasing the speed of rotation. Therefor adsorption rotors can be used monovalent for domestic heating. Investment costs of the heating installation are reduced, because no additional boiler is necessary.

Field of application

Adsorption rotor systems are apt to a wide field of application. The suggested working pair zeolite/water is well suited for heat transformator (Rotmeyer, 1985) and heat pump applications (Maier-Laxhuber, 1983). Zeolite in combination with the refrigerant ammonia allows evaporation temperatures far below zero (Poppendick, 1987). Compact heat exchangers are suitable for liquid as well as for gaseous heat transfer agents. The capacity can be varied in a wide range by the dimensions of the rotor and the number of units. Because of the lightweighted and compact heat exchanger design they are suitable for mobile applications, e. g. air conditioning in cars and buses.

This project is financed by the BMFT Germany

References

Bock J.A., Gundlach D.P., Khoury I.A., Langer G.G.,1987, "Thermal Analysis of Adsorbents in Heat Transfer Systems," Thermochimica Acta, 121.

Kaubek F.,Maier-Laxhuber P., 1983, "Deutsche Patentschrift DE 3342985,".

Knoche K.F., Zerres H., Seitz C.W., 1990, "Dynamic Simulation Of Periodically Operating Absorption Systems," 3rd International Workshop on Research Activities on Advanced Heat Pumps, Graz

Knoche K.F., Seitz C.W. and Zerres H.,1991, "Operating Data of a Directly Fired Absorption Heating Unit with Periodically Operating Absorber," Chemical Engigneering and Technology.

Maier-Laxhuber P.K., 1983, "Sorptionswärmepumpen und Sorptionsspeicher mit dem Stoffpaar Zeolith-H2O," Diss. Universität München.

Niebergall W., "Absorptionskältemaschinen," Plank, R., "Handbuch der Kältetechnik," Band 7, Springer Verlag.

Pohl U.,1986, "Energetische Analyse periodischer Sorptionswärmepumpen," Diss. Universität Hannover.

Pons. M., Guilleminot J.J., 1986, "Design of an Experimental Solar-Powered, Solid-Adsorption Ice Maker," ASME Transactions 332/Vol. 108.

Poppendick K. E.,1987, "Experimentelle Bestimmung der Zustandsgrößen von Zeolith-Ammoniak Systemen," VDI Fortschrittsberichte Nr. 203. Rothmeyer M., 1985, "Wärmetransformation mit dem Stoffpaar Zeolith-Wasser;" Diss. Universität München

Tchernev D. I.; Clinch J. M.,1989, "Closed Cycle Zeolite Regenerative Heat Pump," Solar Engineering, Book No. 467.

STUDY OF THERMOELECTRIC TECHNOLOGY FOR AUTOMOBILE AIR CONDITIONING

B. Mathiprakasam and P. Heenan
Midwest Research Institute
Kansas City, Missouri

V. C. Mei and F. C. Chen
Oak Ridge National Laboratory
Oak Ridge, Tennessee

Abstract

An analytical study was conducted to determine the feasibility of employing thermoelectric (TE) cooling technology in automobile air conditioners. The study addressed two key issues — power requirements and availability of thermoelectric materials. In this paper, a mathematical model was developed to predict the performance of TE air conditioners and to analyze power consumption. Results show that the power required to deliver a cooling capacity of 4 kW (13,80 Btu/h) in a 38°C (100°F) environment will be 9.5 kW electric. Current TE modules suitable for air conditioning are made of bismuth telluride. The element tellurium is expected to be in short supply if TE cooling is widely implemented for auto air conditioning; some options available in this regard were studied and presented in this paper.

Nomenclature

A	=	TE element cross-section area
C_a	=	specific heat of air
C_w	=	specific heat of liquid
COP_c	=	cooling coefficient of performance
H_a	=	product of air-side heat transfer coefficient and area
H_w	=	product of liquid-side heat transfer coefficient and area
h_a	=	enthalpy of air
I	=	electrical current flow
J_a	=	product of air-side mass transfer coefficient and area
k	=	thermal conductivity of a TE couple
l	=	TE element length
L_1	=	characteristic length of liquid channels in the TE heat rejector/cooler
L_2	=	characteristic length of air passage in the TE heat rejector/cooler
M_a	=	air mass flow rate
M_w	=	liquid mass flow rate
N	=	total number of thermoelectric couples
NTU	=	number of transfer units
P	=	power input to the TE system
T_a	=	air temperature
T_c	=	TE module cool-side temperature
T_h	=	TE module hot-side temperature
T_w	=	liquid temperature
W_a	=	air moisture content

Nomenclature (cont)

W_c	=	air moisture content at temperature T_c
X	=	space coordinate in the liquid flow direction
Y	=	space coordinate in the air flow direction
α	=	Seebeck effect of TE couple
ν	=	area/length ratio of TE element
ρ	=	TE couple electric resistivity

Subscripts

A	=	cooler
n	=	TE module negative element
p	=	TE module positive element or pump-side
R	=	heat rejector

Introduction

Increased awareness of the impact of chloroflurocarbons (CFCs) on the global environment has become the impetus in searching for alternative refrigerants and cooling methods for automotive air conditioning. Automotive air conditioning is one industry that heavily uses CFC compounds, and the leakage of CFCs from such air conditioners is substantial compared to that from stationary air conditioners. Therefore, the use of non-CFC air conditioners in automobiles is apparently becoming very important.

There are a number of air conditioning technologies that are not CFC-based. One of them is the well-known thermoelectric (TE) technology. The air conditioners employing this technology use some kind of specialized semiconductors instead of working fluids: thus, the core of a TE air conditioner is a completely solid-state device. In general, TE technology appears to possess characteristics that are favorable for automotive applications. These include (a) a match in power source — dc power is required to operate TE units, and the power system in automobiles is dc; (b) high reliability and easy maintainability — important for consumer products such as automobiles; (c) compact size; (d) quietness; and (e) easy capacity modulation. Factors that are unfavorable to TE technology include (a) relatively low efficiency (TE units' efficiency is lower than that of CFC-based units, yet is very comparable to that of many other non-CFC units) and (b) availability of TE materials to meet the quantity of production required for auto air conditioning.

Program Objectives

The development of TE air conditioners for automobile applications will be a multiphase, multiyear program. This study addressed only the first phase of this effort — a feasibility study. The objectives of this phase of the program were (a) to design a conceptual TE air conditioner suitable for automobiles, (b) to determine a set of design conditions applicable for a typical compact car, (c) to calculate the power requirements to operate the TE air conditioner using the existing computer design tools, and (d) to investigate material availability issues.

The conceptual design of a TE air conditioner included the formulation of a schematic representing various components of the system and the description of the system operation. The determination of a set of design conditions included the review of literature and selection of appropriate values for quantities such as cooling load, ambient temperature, etc. The power requirements portion of the study included the use of the existing computer model to estimate the power required to run the air conditioner. The material availability study included the collection of data on the current rate of production of TE modules and the impact on future production should these modules also be used in automobile air conditioners.

Description of a Conceptual System

The conceptual TE system has four essential components: a dc generator, a blower, a cooler, and a heat rejector. The schematic of the TE air conditioner system is shown in Fig. 1. The dc generator is where the compressor of a conventional freon system would be located. This high-voltage dc generator (voltage depends on the number of TE modules) is driven directly by the engine to supply the necessary dc power to the TE modules. A blower is used, as in the conventional CFC system, to pump the air to be conditioned; this stream of air is called "process air" in this study. The process air is either the returning interior air or a mixture of fresh and interior air. The blower draws process air, pumps it through the cooler, and finally sends it back to the car interior via several vents.

The cooler is located where the evaporator of the current CFC system would be located. This unit is a thermoelectric solid state device having several TE modules. It removes heat and moisture from the process air using dc power, and rejects heat to a transfer fluid to be used in the system. The transfer fluid is a mixture of ethylene glycol and water, which is the same as that used in automobile radiator fluid. The last component is the heat rejector. This unit is located where the condensing unit of the CFC system would be placed. The heat rejector is also a thermoelectric device capable of removing heat from the transfer fluid and rejecting it to the ambient air.

The flow rate of process air and the mix of interior/outside air can be controlled in the same manner as in the existing air conditioning system. The temperature of air exiting the vents, however, controlled by regulating the dc power applied to the cooler and/or the heat rejector.

Mathematical Model

The system schematic is shown in Fig. 1. The cooling system is composed of two cross-flow thermoelectric - liquid (water/ethylene-glycol mixture) heat exchangers, a heat rejector, and a cooler (heat absorber). A liquid circulating pump is also included. Based on the energy balance of the heat transfer medium, air, and liquid (water/ethylene-glycol mixture), the following mathematical model can be formed.

Heat Rejector:

- Liquid side —

$$-M_W C_{WR} L_{1R} \frac{\partial T_{WR}}{\partial X} = H_{WR}(T_{WR} - T_{CR}) \text{, and} \qquad (1)$$

$$-M_W C_{WR} L_{1R} \frac{\partial T_{WR}}{\partial X} = N \left[\alpha T_{CR} I_R - \frac{0.5 I_R^2 \rho}{\nu} - k\nu (T_{hR} - T_{CR}) \right] , \qquad (2)$$

where

$$k\nu = k_p A_p / l_p + k_n A_n / l_n \text{, and}$$

$$\rho/\nu = \rho_p l_p / A_p + \rho_n l_n / A_n .$$

- Air side —

$$M_{aR} C_{aR} L_{2R} \frac{\partial T_{aR}}{\partial Y} = H_{aR}(T_{hR} - T_{aR}) \text{, and} \qquad (3)$$

$$M_{aR} C_{aR} L_{2R} \frac{\partial T_{aR}}{\partial Y} = \left[\alpha T_{hR} I_R + \frac{0.5 I_R^2 \rho_R}{\nu} - k\nu (T_{hR} - T_{CR}) \right] . \qquad (4)$$

Cooler (heat absorber):

- Water side —

$$M_W C_{WA} L_{1A} \frac{\partial T_{WA}}{\partial X} = H_{WA}(T_{hA} - T_{WA}) \text{, and} \qquad (5)$$

$$M_W C_{WA} L_{1A} \frac{\partial T_{WA}}{\partial X} = N \left[\alpha T_{hA} I_A + \frac{0.5 I_A^2 \rho}{\nu} - k\nu (T_{hA} - T_{CA}) \right] . \qquad (6)$$

- Air side —

$$- M_{aA} C_{aA} L_{2A} \frac{\partial T_{aA}}{\partial Y} = H_{aA}(T_{aA} - T_{cA}) , \qquad (7)$$

$$-M_{aA} L_{2A} \frac{\partial h_{aA}}{\partial Y} = n \left[\alpha T_{cA} I_A - \frac{0.5 I_A^2 \rho_A}{\nu} - k\nu (T_{hA} - T_{cA}) \right] \text{, and} \qquad (8)$$

$$-M_{aA} L_{2A} \frac{\partial W_{aA}}{\partial Y} = J_{aA}(W_{aA} - W_{cA}) . \qquad (9)$$

The system cooling COPs can be calculated by the following equation:

$$COP_c = \frac{\text{Cooling Capacity}}{P_A + P_R} . \qquad (10)$$

The computer code was based on a previous study (Mei et al, 1989) modified for this application.

Realistic TE material properties, as shown below, were used for the simulation.

Seebeck Coefficient

α = 1.8 x (2T + 1985) 10^{-7} volt/°C.

Electrical Resistance

ρ = (6T + 1735) 10^{-6} ohm cm.

Thermal Conductivity

k = 0.0324 watt/cm/°C.

where

$$T = \frac{(T_h + T_c)}{2}$$

The above TE property figures are for those typical off-the-shelf TE modules, including the Semiconductors, copper straps, and ceramic substrates.

Design Conditions

Based on a literature review of the cooling requirements of a typical automobile, a set of design conditions were selected to estimate the power requirements of the TE air conditioner.

Cooling Load. The paper published by Ruth [1975] contains the cooling requirement data for various car types, ambient weather conditions, driving conditions. His data, shown in Table 1, were considered for selecting the design cooling load. QA compact car was also chosen for this analysis. We considered the worst-case cooling load for ambient conditions of up to 38° C (100° F); the cooling needs at 43° C (110° F) ambient, were not considered for design since it was felt that the probability of an air conditioner operating under such ambient conditions would be extremely low. According to Ruth's data, a cooling requirement of 4.01 kW (13,680 Btu/h) is the worst case at 38° C (100° F) ambient; therefore, it was decided to choose a design cooling load of 4.01 kW (13680 Btu/h) at 38° C (100° F) ambient. The cooling requirement at 32° C (90° F) ambient is up to 4.17 kW (14,220 Btu/h) because TE efficiency increases at a lower ambient temperature, which is more than the chosen design value; however, the TE air conditioner designed to provide 4.01 kW (13,680 Btu/h) at 38° C (100° F) ambient will provide more than 4.17 kW (14,220 Btu/h) in a 32° C (90° F) environment.

State of interior air. Several investigators in the area of car air conditioning have used a design state of 25° C (77° F) at 60% relative humidity (RH) as the design value. This value seems very appropriate from the standpoint of the occupants' comfort. Therefore we decided to use this state as our design state of interior air.

State of process air at stage 1. Again per Ruth's data, a cooling requirement of 4.01 kW (13,680 Btu/h) applies when the system is operating under a 100% recirculating mode. Under this condition, the state of process air at stage 1 (see Fig. 1) should be the same as the state of interior air [25° C (77° F) and 60% RH]. The humidity ratio of this stream is 0.0119 gr water per gr dry air (0.0119 lb water per lb dry air).

State of Process air at stage 2. This is the state of the air at the entrance to the cooler. Because of the blower heat and the heat transmitted from the engine and picked up by the panel, the temperature at stage 2 is expected to be about 8.3° C (15° F) more than at stage 1. As such, we chose a design temperature at stage 2 of 25 + 8.3 = 33.3° C (77 + 15 = 92° F). The humidity ratio at stage 2, however, will be the same as stage 1. The RH of air at 33.3° C (92° F) dry bulb and 0.0119 humidity ratio is 37%; thus, the state of process air at stage 2 will be 33.3° C (92° F) and 37% RH. The enthalpy of process air at stage 2 is 81.97 kJ/kg (35.24 Btu/lb).

Process air flow rate. For current automotive air conditioners, the design man flow rate of process air varies from 386 to 567 kg/h (850 to 1250 lb/h)]. For our design, we have chosen 454 kg/h (1000 lb/h), which seems reasonable for a compact car.

State of process air at stage 3. Having known the total cooling load, the enthalpy of air at stage 1, and the mass flow rate, we calculated the enthalpy of air at stage 3, using

Cooling = Mass flow rate (enthalpy at stage 2 - enthalpy at stage 3)

The enthalpy of process air at stage 3 was thus found to be 50.14 kJ/kg (21.56 Btu/lb). From psychrometrics, this relates to an air state of 11° C (52° F) and 100% RH.

State of ambient air at stage 6. Just as we did while selecting the cooling requirements, we chose an ambient temperature of 38° C (100° F) as the design value at stage 6.

Transfer fluid temperature change is. The fluid used as the transfer fluid is the ethylene-glycol water mixture. This fluid is heated while passing through the cooler and cooled while passing through the heat rejector. Since both of these units will have a cross-flow configuration, first law of thermodynamics dictates that the temperature increase in the cooler (which is also equal to the temperature drop in the heat rejector, assuming no thermal losses in the line) should be as small as possible. Our estimate is that this increase in temperature should be limited to about 3 to 5.5° C (5 to 10° F). An increase of less than 3° C (5° F) will require an excessively large heat exchanger, and an increase of more that 5.5° C (10° F will make the TE system inefficient.

Volumetric flow rate of transfer fluid. The volumetric flow rate of transfer fluid depends on the amount of heat removed from the cooler and the temperature rise in the cooler. For the purpose of power analysis, we used three different flow rate values. 1.51, 1.89 and 2.27 m^3/h (400, 500, and 600 gal/hr). We estimate that we can achieve a COP of 1.5 in the cooler based on our past experience; this means that the heat to be removed in the cooler will be 4.01 kW + 2.67 kW = 6.68 kW (13,680 + 13,680/1.5 = 22,800 Btu/h). The heat capacity of the 50/50 ethylene-glycol/water mixture is about 0.989 kW/m^3/° C (7.11 Btu/gal/° F). At 1.89 m^3/h (500 gal/h) flow rate, the temperature rise in the cooler will be 6.68/0.989/1.89 = 3.6° C (22,800/7.11/500 = 6.4° F), which is well within the acceptable range.

Number of transfer units (NTU) values. NTU is defined as the ratio of the total heat transfer rate (in Btu/h/° F) on any side of a heat transfer device to the heat capacity of the fluid (in Btu/h/° F). At this time, we estimate that we can design the cooler such that the NTU on the process air side will be about 6.0 and that on the transfer fluid side will be about 1.0. We also estimate that we can design the heat rejector such that the NTU on the transfer fluid side will be about 1.0 and that on the ambient air side about 0.9.

Power Analysis

A computer simulation of the TE air conditioner system was performed using the design conditions. The output from the computer included the cooler COP, heat rejector COP, overall COP, temperatures of transfer fluid at stages 4 and 5, and other information pertinent to the selection of TE modules.

The cooler COP is defined as the ratio between the useful cooling delivered to the process air and the electric power applied to the cooler. The heat rejector COP is defined as the ratio between the useful cooling delivered to the transfer fluid and the electric power applied to the heat rejector. The overall COP is defined as the ratio between the useful cooling delivered to the process air and the total electric power applied to both the cooler and the heat rejector.

Table 2 contains the values of different COPs mentioned above. Figure 2 shows the values of overall COP in graphical form. As can be seen from the table values, for a given flow rate of transfer fluid, the cooler COP decreases as the transfer fluid temperature at the cooler inlet is increased; at the same time, the rejector's COP increases. The overall COP attains a maximum value at some optimum value of transfer fluid temperature. In general, this optimum temperature is around 24°C (75°F).

For a given transfer fluid temperature at the cooler inlet, the cooler COP, rejector COP, and overall COP all increase as the flow rate of transfer fluid is increased. This suggests that the flow rate of the transfer fluid should be as high as is practical. This will reduce the temperature change of the transfer fluid in the cooler (and in the rejector). Theoretically, a fluid with zero temperature change will offer the highest COP values. From this perspective, a phase change fluid (which will evaporate in the cooler and condense in the rejector) might be a better candidate than the chosen ethylene glycol.

At a flow rate of 2.27 m^3/h (600 gal/h) and 24°C (75°F), the analysis shows that an overall COP of 0.424 is achievable employing the TE technology. Based on the limited data available on conventional air conditioning systems, we project the COP of such systems to be in the area of 1.0. As such, the COP of air conditioners using TE technology appears to be low. However, we feel that this COP will still be competitive with other non-CFC-based systems. For a design cooling load of 4.01 kW (13680 Btu/h) at 38°C (100°F) and the associated COP of 0.4241, the power consumption will be 9.451 kW (12.67 hp). In a car with an engine power of, say, 120 to 130 hp, this represents a 10% power draw (CFC system power draw is about 7%).

Materials Availability

In order to produce TE air conditioners in large quantities, it is necessary to evaluate the availability of all of the materials that are required to build these systems. A basic unit consists of thermoelectric modules, aluminum sheets and tubes, and various aluminum fin materials. Since the aluminum parts are similar to those or existing air conditioning components and since aluminum is unlimited in supply for purposes of this evaluation, we need not consider these components. The TE modules, however, have limitations that must be explored in this evaluation.

A number of issues can influence the production ability of TE module fabricators. These key issues include, but are not limited to, raw materials availability, refined materials availability, materials cost, existing module production capacity, and the ability to increase module production capacity. These issues will be presented and discussed below. Much of this information was obtained from Melcor, Inc. (Mathiprakasam, 1991).

Four materials are used to make up the basic TE components. Bismuth and tellurium are the primary ingredients, and small portions of selenium and antimony are added in different proportions to create the N and P polarity of the resulting semiconductor materials. Of these four materials, bismuth, selenium, and antimony are widely available from multiple sources with considerable excess capacity potential at no increase in price. Only tellurium has a finite limit on availability with some cost dependency.

Tellurium is not practical to mine directly and is a by-product of other mining operations. The primary source of tellurium is the refining waste of copper, although tellurium is also present and recoverable from lead, silver, and gold refining wastes. Only copper mining is of significant quantity to be used as a practical source for tellurium. Additional demands for tellurium above that available from these sources will not affect these mining activities. The availability of tellurium is thus tightly keyed to copper industry demands.

The total production of tellurium from these sources amounts to about 363,000 kg (800,000 lb) of usable refined material per year. This amount has been extremely stable, and level of production is not affected significantly by anything known to date. TE and other semiconductor producers consume about 10% of this production with the remainder of the production consumed by steel and other metals industries as well as other minor chemical processes. The 10% consumed for semiconductors is largely for TE material production. The tellurium used for TE production must be further refined to increase its purity. The other 90% is consumed as is from the initial refining process.

Actual modules produced today widely differ in size and type and use different volumes of materials. To provide numerical estimates of module production capacity, it is assumed that a 9.7-cm^2 (1.5-$in.^2$), 127-couple module is standard for potential use in automotive air conditioning. It is estimated that about 50 of these standard modules are made from about 1 lb of tellurium assuming large amounts of nonrefined copper wastes are generated and accepted. If more care is taken in production and the waste is further refined, it is estimated that this rate could increase to 100 standard modules per pound of tellurium. This increase in modules per pound will result in small but noticeable increases in cost for the end product as a result of increased labor and tellurium refining costs.

The standard module is smaller than the average TE module fabricated and sold today. Because of this, estimates for production rates of these standard modules will seem large when compared with actual production rates for the same consumption rate of tellurium. If we assume that the most efficient method of obtaining tellurium is employed to maximize the modules produced per pound, we can assume that 8 million of the standard modules can be fabricated with the same quantity of tellurium that produces 1.2 million various actual modules using today's methods.

Although the tellurium is available to fabricate as many as 8 million TE modules, the fabricator would need about 6 months to increase production to about 4 million modules a year by hiring, training and adding additional shifts to run the existing facility continuously. At this point, new facilities would have to be built to increase production, which could take as long as another year and would require stable long-term commitments from the end user to make it practical for the fabricator. All of this expansion would not influence TE module price significantly. In fact, if this demand were real, the fabricator would move more quickly to automate the plant, which would increase production with no immediate increase in final TE module cost and no long-term reductions in TE module costs.

If the maximum number of TE modules were limited to the 10% of available tellurium, the number of automotive TE air conditioning units that could be built would be restricted. At this time, it appears that about 400 standard TE modules would be required per vehicle if the TE air conditioner were fabricated to provide the same level of cooling as current vapor-cycle AC units. If all 8 million modules fabricated were exclusively available for this effort, about 20,000 vehicles could be fitted with TE air conditioners per year.

The TE market, however, is not limited to the 10% available tellurium: almost all other tellurium users to date use the least refined product, and other materials could easily be used as a substitute for tellurium. Because extra refining is required for the tellurium used in TE production, it is a value-added process to the refiner. Since refiners make more profit selling to TE producers, TE is a preferred market. Even if TE producers wanted almost all of the tellurium, the refiners would be more than willing to accommodate them. If existing users then still pursued buying tellurium, the price could be driven up, but it would be limited to the cost of substituting another material for their processes. Even if the price of tellurium doubled, the price of TE modules would only be increased by about 25%. Tellurium availability, therefore appears to be a limited problem with easy solutions.

A practical assumption about the maximum amount of tellurium available for TE air conditioner production will allow for about 80% of the mined tellurium to be dedicated. The 10% currently used will still be used for TE module production, with 10% being left for other non-TE uses. Based upon these assumptions, 290,000 kg (640,000 lb) of tellurium will be available exclusively for producing TE modules for automotive AC application. This means that 64 million modules can be fabricated, which translates to fitting 160,000 vehicles with TE air conditioners.

Finally, fitting TE air conditioners on more than 160,000 vehicles would seem challenging since the tellurium needed is keyed to a potentially required increase in copper production. If this scenario were a reality, tellurium production capacity might be increased, but innovation here would only be possible if the profits were sufficiently high to pursue alternate sources outside of refined mining wastes. At this point, costs for additional tellurium could escalate dramatically. It is not possible to estimate these potential cost impacts at this time with currently available information.

Conclusions

The first phase of the program on the development of TE air conditioners for automobiles — a study of the feasibility of employing TE technology — was completed. Two key issues were addressed in this phase. It was found that a TE air conditioner delivering a total cooling of 4.01 kW (13,680 Btu/h) will require a total dc power of about 9.5 kW, which is equivalent to a cooling COP of 0.42. This power also represents about 10% of the total power of the engine. The power consumption of TE air conditioners seems higher than that of conventional CFC-based air conditioning systems. On the material side, it was found that one of the elements used to make TE modules — tellurium — will have availability problems. Current production of TE modules can be increased by about 10 times without a price impact. Beyond that point, the cost of TE modules will be dependent on the quantity.

Acknowledgments

This work was a part of the Transportation Environmental Control Activities at Oak Ridge National Laboratory sponsored by the CFC Reduction Program of the Office of Transportation Technologies, U.S. Department of Energy, under Contract No. DE-AC05-84OR21400 with Martin Marietta Energy Systems, Inc.

References

Mathiprakasam, B., Midwest Research Institute, Kansas City, MO, personal communication with Melcor, Inc., Trenton, NJ, 1991.

Mei, V. C., Chen, F. C., and Mathiprakasam, B., "Comparison of Thermoelectric and Vapor Cycle Technologies for Ground Water Heat Pump Application," ASME J of Solar Eng. V (III), pp. 353–357, Nov. 1989.

Ruth, D.W., "Simulation Modeling of Automobile Comfort Cooling Requirements," ASHRAE Journal, May (1975), pp. 53-55.

TABLE 1 - AUTOMOTIVE COOLING REQUIREMENTS

Car Type	Ambient Condition °C (°F) / RH	City Driving (Cool Down) No Outside Air kW (Btu/h)	City Driving 30 mph 100% Outside Air kW (Btu/h)	Highway Driving 60 mph 100% Outside Air kW (Btu/h)
Subcompact	32 (90) / 50%	3.59 (12250)	3.49 (11910)	3.77 (12850)
	38 (100) / 20%	3.47 (11830)	3.20 (10930)	3.41 (11640)
	43 (110) / 5%	3.86 (13170)	3.79 (12950)	4.08 (13940)
Compact	32 (90) / 50%	4.14 (14140)	4.17 (14220)	4.14 (14120)
	38 (100) / 20%	4.01 (13680)	3.76 (12840)	3.73 (12730)
	43 (110) / 5%	4.42 (15100)	4.51 (15380)	4.48 (15280)
Standard	32 (90) / 50%	5.06 (17270)	5.16 (17620)	5.13 (17520)
	38 (100) / 20%	4.91 (16770)	4.64 (15830)	4.61 (15730)
	43 (110) / 5%	5.37 (18320)	5.55 (18950)	5.52 (18850)

TABLE 2 - COEFFICIENT OF PERFORMANCE

Transfer Fluid		Cooler	Rejector	Overall
Flow Rate m^3/h (gal/h)	Cooler Inlet Temperature °C (°F)	COP	COP	COP
	18 (65)	3.318	0.460	0.3197
	21 (70)	2.363	0.546	0.3300
1.51 (400)	24 (75)	1.781	0.633	0.3303
	27 (80)	1.389	0.721	0.3222
	29 (85)	1.108	0.806	0.3064
	18 (65)	3.836	0.516	0.3696
	21 (70)	2.686	0.613	0.3830
1.89 (500)	24 (75)	2.009	0.718	0.3872
	27 (80)	1.564	0.833	0.3836
	29 (85)	1.250	0.958	0.3732
	18 (65)	4.234	0.548	0.4010
	21 (70)	2.925	0.653	0.4171
2.27 (600)	24 (75)	2.173	0.769	0.4241
	27 (80)	1.687	0.901	0.4236
	29 (85)	1.348	1.050	0.4164

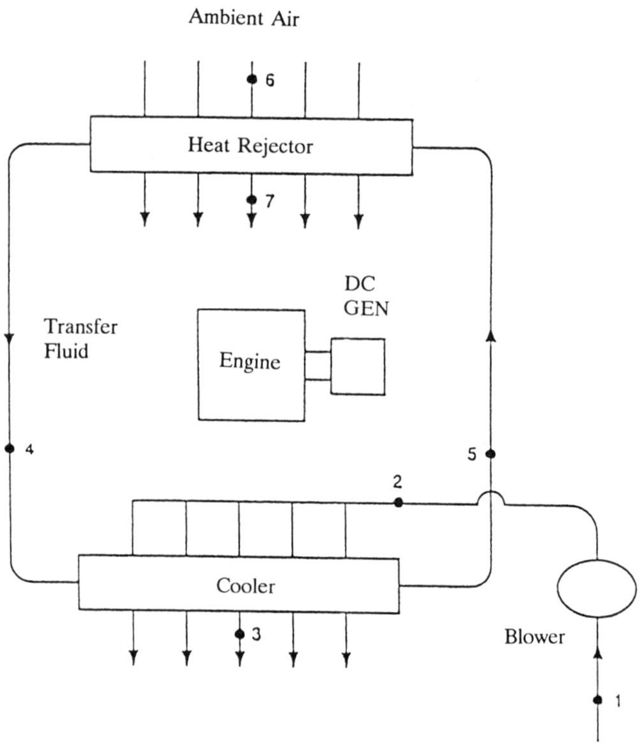

Fig. 1. Schematic of a TE automobile air conditioning system.

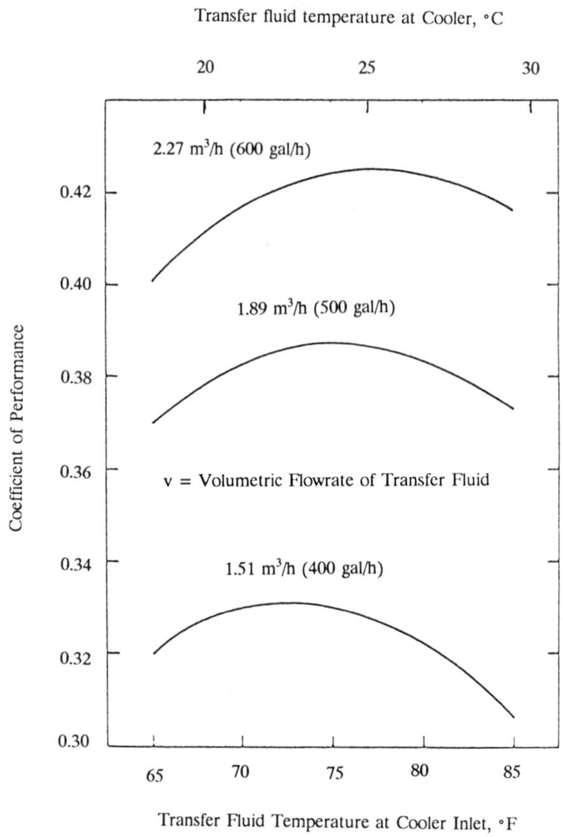

Fig. 2. System performance as a function of transfer fluid cooler inlet temperature.

PRACTICAL PROBLEMS IN THE IMPLEMENTATION OF HEAT PUMP IN THE SYNTHETIC RUBBER INDUSTRY

Deepak Shukla and Rajiv Singh
TENSA Services, Incorporated
Houston, Texas

Dennis Perkinson
American Synthetic Rubber Corporation
Louisville, Kentucky

ABSTRACT

Heat Pump applications were examined for American Synthetic Rubber Corporation's Polybutadiene Plant. The preliminary studies indicated promising economics for two Mechanical Vapor Recompression (MVR) heat pumps. One heat pump has a potential savings of 15.4 MMBtu/Hr, with a payback of ten (10) months. The other MVR heat pump takes advantage of the overhead vapors from the stripping section. By compressing the overhead vapors from the strippers with the MVR heat pump, the heat of condensation of the vapors can be used to generate steam that is needed elsewhere in the plant. This application has a potential savings of over 50 MMBtu. However, these vapors contain rubber particles which deposit inside the condenser as "crumb".

This paper presents the details of the heat pump and examines the various options of solving the "crumb" problem. The first option examined considered mechanical cleaning devices such as filters, strainers, on-line cleaning brushes and baskets. The second option evaluated the feasibility of generating clean steam against the dirty condensing vapor. The clean steam would then be utilized in the heat pump.

1.0 INTRODUCTION

Over the recent years there is an increasing awareness for energy conservation. One of the areas where significant effort has been made is in the area of heat integration (1-4). It is felt that through better heat integration and the use of such energy saving applications as heat pumping, the U.S industry can save significant amount of steam consumption. These energy savings will not only translate to savings in cost of product but will also reduce the amount of coal or natural gas being fired into the boilers, thus reducing the emissions of CO_2, NO_x and SO_x into the atmosphere(5).

Over the past several years, TENSA Services has been working on a U.S. Department of Energy (U.S.D.O.E.) Office of Industrial Technologies funded project for the application of industrial heat pumps (6, 7). A multi-phased approach was adopted to demonstrate the applicability of industrial heat pumps in various process industries. This paper describes the results of one of the studies performed by Litwin Engineers and TENSA Services. The study included three parts - pinch analysis, semi-open cycle heat pump and closed cycle hea tpump. The first two functions were performed by TENSA Services and the third function was performed by Litwin Engineers. The main tasks performed by TENSA Services are described in the following sections.

2.0 TECHNOLOGY FOR PROPER PLACEMENT OF INDUSTRIAL HEAT PUMPS

The methodology developed for the proper placement of industrial heat pumps uses the pinch technology process integration method (1-4). It may be worthwhile to point out that some previous attempts at placing industrial heat pumps in processes resulted in no savings and in some cases increased the total energy consumption by the plant because of a lack of understanding of the impact of the heat pump on the overall process. Therefore, use of pinch technology attempts to avoid such bad experiences by providing a systematic way of examining all processes. This technology is based on the fact that most industrial processes have heating and cooling requirements, which can be divided into two parts:

- A high temperature part which acts as a net acceptor of heat, called "heat sink".

- A lower temperature region which has an excess of heat and acts as a heat source.

The temperature which separates these two regions is called the 'pinch' temperature. In a properly integrated process, there is no heat transfer from above the pinch to below the pinch. For the heat pump to be properly placed according to pinch principles, it must take waste heat from below the pinch and deliver it above the pinch. This can be best examined by the Grand Composite Curve (GCC) (Figure 1), which represents the composite of all the streams between the interval temperatures. For the heat pump to be viable and properly placed, there should be a long flat region below the pinch, which usually corresponds to a condensing stream. Short of the stream condensing, this stream can be compressed and then allowed to condense at a higher temperature to deliver the heating duty above the pinch. This heat pump is called a Mechanical Vapor Recompression (MVR) heat pump with a condenser (Figure 2). In another variation of the MVR heat pump, the condensing vapor is first used to deliver heat at a lower temperature, and the stream to which heat is delivered, is then compressed to a higher pressure and temperature.

It may be emphasized that in order to find the true economic payback for the heat pump the heat integration option is examined first, because it usually is the more economic option, as it requires additional heat exchange area which is relatively inexpensive. However, through heat integration, one can only save a portion of the waste heat. A heat pump option offers a slightly more expensive option as compared to heat integration (primarily because of the compressor cost), however the waste heat saved can be significant.

3.0 PROCESS DESCRIPTION

The American Synthetic Rubber Plant in Louisville, Kentucky produces polybutadiene and styrene butadiene rubber (SBR). The scope of the study described in this paper was confined to two trains of the polybutadiene rubber. The process described here is similar for most synthetic rubber processes. Polybutadiene is produced by the polymerization of butadiene monomer in the presence of toluene solvent. The remainder of the process involves concentrating the rubber in stages and removing toluene vapor, which is then dried and recycled. Toluene removal is performed by a stripping process, where steam is directly injected into the process. Steam injection serves two main purposes - it provides the necessary latent heat for toluene to evaporate, and also lowers the partial pressure at which toluene can be removed. The steam-toluene mixture is then condensed and sent to the toluene drying section.

4.0 HEAT INTEGRATION OPTION

The process was examined for heat integration options, and the results indicate that there is a possibility of saving 9.7 MMBtu/Hr in steam for an investment in area of 8500 sq. ft. over the existing network. The investment will payback within 2 years. The details of the heat exchanger network can be found in the Phase I report (7).

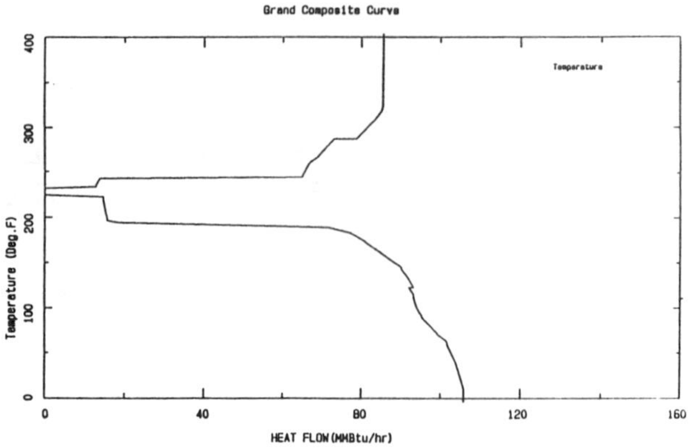

FIGURE 1

Grand Composite Curve (GCC) for the Polybutadiene Rubber Process

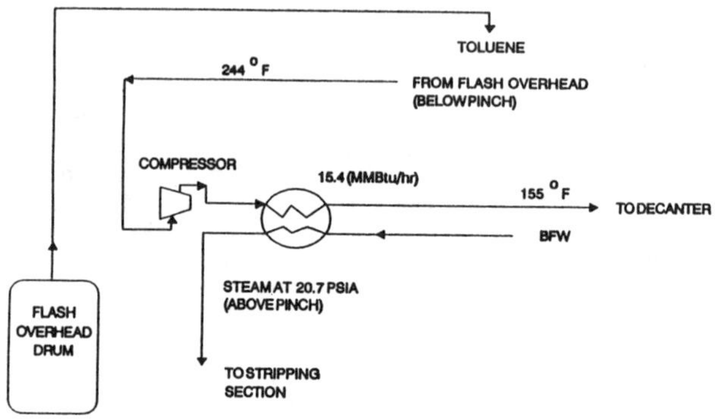

FIGURE 2

Schematic Representation of MVR Heat Pump #1

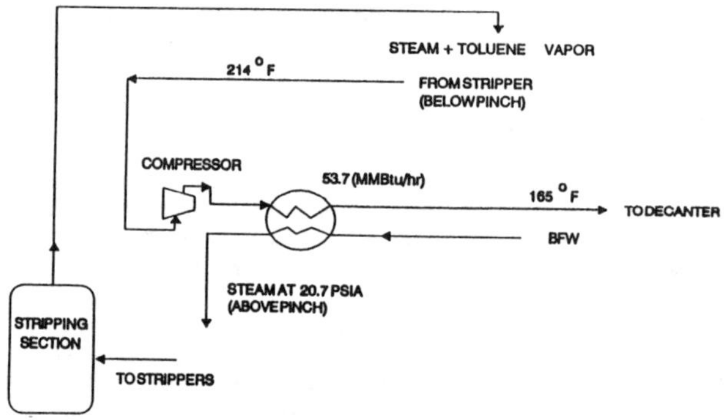

FIGURE 3

Schematic Representation of MVR Heat Pump #2

5.0 HEAT PUMP APPLICATION

Two industrial heat pump applications were identified in this phase of the project, based on the pinch principles described earlier. One of the heat pumps involves utilizing the flash overhead stream from the concentration section, which is composed of 99% toluene. This vapor is presently being condensed and recycled back into the reactor. The other heat pump utilizes steam and toluene vapor from the strippers, and has a condensing duty of 53.1 MMBtu/Hr, which is condensed in stripper overhead condensers. The condensed steam-toluene mixture is the decanted and sent to the toluene drying section, to be recycled.

5.1 Heat Pump #1

The first heat pump opportunity identified in this study utilized the flash overhead vapor from the concentration section. This vapor is currently condensed in the flash overhead condenser. The proposed heat pump will compress this vapor and condense it against boiler feed water to generate steam at 20.7 psia, as shown in Figure 2. The design of the heat pump was optimized for the condenser temperature. The details of the optimized design are presented in Table 1.

As is apparent from the table the economics for this heat pump are very attractive. The economics are based on a steam price of $2.84/MMBtu and an electricity price of $0.04/KWH.

TABLE 1

SUMMARY OF HEAT PUMP #1
(FLASH OVERHEAD VAPOR) DESIGN

Compressor Size (Brake Horse Power)	244
Compression Ratio	1.67
Temperature Lift (°F)	20
Heat Pump COP	27.2
Condenser Area (ft.^{2})	5572
Condenser Cost @ $40/ ft.^{2}	222,902
Total Capital Cost ($)	305,030
Hot Utility Saved (MMBtu/Hr)	15.4
Compression Cost ($/Yr)	60,023
Net Total Utility Savings ($/Yr)	353,298
Payback Period (Yrs)	0.9

5.2 Heat Pump #2

This MVR heat pump utilizes the stripper overhead vapor to generate steam for direct injection into the process. This vapor is composed of steam-toluene mixture, which is currently condensed in the stripper condenser at 214°F, and the after decanting out the water, is sent to the toluene drying section. The proposed heat pump will compress the overhead vapors from the strippers and condense it against boiler feed water to generate steam at 20.7 psia (Figure 3). As described before, there are higher savings in steam using this heat pump, however, the capital investment in this heat pump is significantly higher. The economics for this heat pump is presented in Table 2.

TABLE 2

SUMMARY OF HEAT PUMP #2
(STRIPPER OVERHEAD VAPOR) DESIGN

Compressor Size (Brake Horse Power)	2941
Compression Ratio	4.1
Temperature Lift (°F)	50
Heat Pump COP	10.5
Condenser Area (ft.^{2})	17,694
Condenser Cost @ $40/ ft.^{2}	705,962
Total Capital Cost ($)	1,310,119
Hot Utility Saved (MMBtu/Hr)	53.7
Compression Cost ($/Yr)	724,306
Net Total Utility Savings ($/Yr)	650,190
Payback Period (Yrs)	2.0

As is apparent from the economics there are significant savings in steam through this heat pump, despite the heavy capital expenditure on the compressor. The combined savings in energy through the two heat pumps would account for over 50% of the total steam consumption by this facility.

6.0 COGENERATION

The option of generating steam at high pressure and then letting it down through a back-pressure turbine was also examined by ASRC (Figure 4). Although the economics presented in this paper are very preliminary in nature, they serve to illustrate a comparison between the heat pump and cogeneration. The total steam demand for the ASRC facility is 170,000 lbs /hr. In this option it is proposed to set up set up three boilers to generate steam at 350 psia which can be let down to 200 psia through a back pressure turbine along with a 3 MW generator. The economics for the cogeneration option is shown in Table 3.

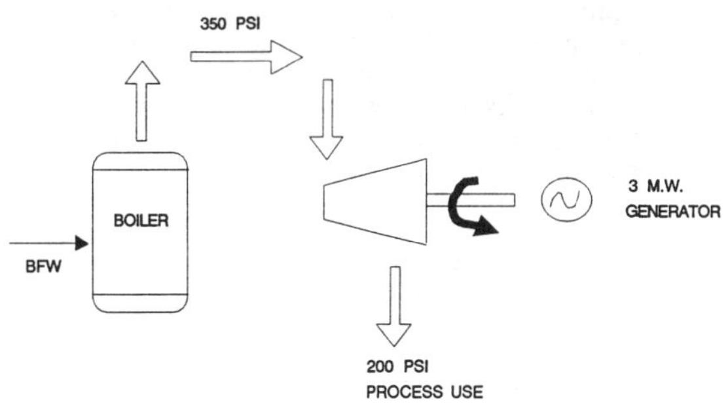

FIGURE 4

Schematic Representation of Cogeneration Option

TABLE 3

COGENERATION ECONOMICS SUMMARY

Present Steam Demand (Lbs/Hr)	170,000
Proposal to set-up 3 boiler total capacity (Lbs/Hr)	350,000
Total Cost Boiler and Infrastructure	20,000,000
Cost of 3 MW Generator and turbine set	1,060,000
Power generated at 85% efficiency	2,260 KW
Annual Credit for Power Generation	746,342
Simple Payback	1.4 Yrs

7.0 INSTALLATION OF HEAT PUMPS

Having performed the design for the heat pump and an economic analysis, one has to examine the details of the process to identify any constraints. There are several problems one encounters, some of which are described in this section.

7.1 Problems with Installation of Heat Pumps

a. Heat Pump #1

Heat pump #1 utilizes toluene vapor as the working fluid, which has been widely used in compressor applications. This vapor is relatively clean and can be compressed to the desired pressure with relative ease. The design and economics for this heat pump are being reviewed before the quotations from the vendor can be obtained.

b. Heat Pump #2

The overhead vapor from the strippers is a mixture of steam (approx. 35% by wt.) and toluene (approx. 65% by wt.). There are two trains from the stripping section, each with a flow of about 70,000 lbs/Hr of vapor. This heat pump has toluene and steam vapor as the working fluid. The stripper overhead vapor also contains small particles of rubber called "crumb" which presently collect on the walls of the pipe and require cleaning during the annual shutdown. Moreover, "crumb", which is about 5/8" in size, can damage the compressor if they get into the compressor suction.

A number of options were examined to overcome this problem:

1) Use of a Knock-Out Drum

Prior to entering the compressor, the vapor stream will have to go through a knock-out drum which will remove any particles or condensed vapors larger than 100 mm in size.

2) Baffle Type Impingement

As the name suggests, the impingement separator allows the particles to be removed after striking some type of surface. This action is better accomplished in pressure systems, where pressure drops can be taken as a result of the turbulence. Particle removal in streamlined flow is less efficient than in turbulent flow, and may not be effective if the path is not sufficiently baffled. Since the "crumb" particles are fairly large in size 5/8", this type of a separator may work, however it introduces fairly large pressure drops, which may make it unsuitable.

3) Centrifugal Separators

Centrifugal separators utilize centrifugal action for the separation of materials of different densities and phases. They are built in a.) stationary of b.) rotary types. They are usually fairly efficient, however they introduce large pressure drop in the line due to loss of velocity and due to constriction.

4) Use of Filters

There are several on-line filter available to remove solid particles from vapors. One class includes a large bag filter. There are several variations of this type of filters depending on the method of cleaning. The pulse-jet cleaning uses a controlled blast of compressed air from the primary into a secondary nozzle, which is magnified by sudden air being drawn into the bag. The sudden release of air causes the bag to expand fully, throwing the collected particles from the outer surface. The dislodged particles fall into a collection hopper. Another variation of this device uses reverse air cleaning to clean the filter.

Also being considered is a continuous type filter, where the filtering surface is continuously moving to offer a new surface to the vapor.

This type of filter seems to be most likely suited for the "crumb" problem since the facility already has uses a basket filter. However the smaller particles are able to escape through it into the condenser. This provides a reliable and convenient way of removing the particles.

5) Use of Liquid Scrubbers

One of the options being considered is to use water in equilibrium with steam, to scrub the small particles of rubber. The water then passes through a filter and is recycled.

6) Use of Evaporator Type Heat Pump

If the options described above do not work then the option of using evaporator type of heat pump (Section 2) will be considered. The steam-toluene vapor will be condensed against boiler feed water to generate low pressure steam, which will then be compressed in the compressor to the desired pressure. This option has one drawback in the fact that the overhead vapors are at fairly low temperatures (205°F), and to generate steam at this temperature is difficult.

The choice of the method will depend on a number of factors such as ease of operability. Currently, TENSA Services is working with ASRC on determining the size of the particles which are able to escape through the filter and the density of these particles.

CONCLUSION

Two heat pump applications were identified, along with heat integration opportunities for ASRC's Polybutadiene plant. There is a potential savings of 50% in the overall steam consumption, which translates to an overall savings of over 1 million dollars annually.

ACKNOWLEDGEMENTS

This work was accomplished with financial support from the U.S. Department of Energy (DOE). The authors would like to thank Mr. Paul E. Scheihing, Mr. R.N. Chappell of DOE, Dr. Stephen Priebe of EG&G, Idaho and Dr. Gene Trott of ASRC, for their support throughout the project.

REFERENCES

1. Ranade, S.M., E. Hindmarsh, and D. Boland, 1987, "Industrial Heat Pumps: New Insights on Their Integration in Total Sites", 3rd Int. Syp. on Large Scale Applications of Heat Pumps, Oxford, England.

2. Ranade, S.M. and Y.T. Chao, "Industrial Heat Pump: Where and When?" Hydrocarbon Processing, October 1990.

3. Chappell, R.N. et.al., "Industrial Heat Pumps - Types and Cost," DE-AC07-76ID7670.

4. P. Tripathi and Y.T. Chao, " Preliminary Screening of Processes for Industrial Heat Pump Applications," IEA Newsletter, Vol. 8, No. 3, September 1990.

5. Paul E. Scheihing and Stephen Priebe, Proceedings of the "Global Pollution Prevention '91 International Conference and Exhibition, Washington, DC; April 3-5.

6. DOE Report DE87009626 "Optimum Heat Pump Placement in Industrial Processes," prepared by TENSA Services for the U.S. Department of Energy (1987).

7. Litwin Engineers and Constructors/TENSA Services Inc., "New Industrial Heat Pump Application to Synthetic Rubber Production - American Synthetic Rubber Corp.," DOE/ID/12791-1.

HIGH-TEMPERATURE ABSORPTION HEAT PUMPS IN INDUSTRY

Lawrence A. Howe and Donald C. Erickson
Energy Concepts Company
Annapolis, Maryland

ABSTRACT

Absorption heat pumps (AHPs) have been applied to such industrial processes as rubber production and alcohol distillation. A new aqueous alkali metal nitrate absorption working fluid extends the upper temperature limit of AHPs to 260°C (500°F). A computer program was developed to estimate the capital cost of industrial AHPs. The program also determines the optimum size of the AHP components to minimize either the simple payback or the internal rate of return. Applications in petroleum refining and steam compression are discussed.

KEY WORDS

Absorption; Heat Pumps; Industrial; High Temperature; Alkitrate; Nitrates

INTRODUCTION

Recent developments in process integration have shed light on how best to integrate heat pumps into a process (Priebe and Scheihing, 1991). Process integration is a systematic method of analyzing any industrial process to find cost and energy savings (Linnhoff et al., 1982). One valuable aspect of process integration is that it determines the minimum possible amount of heating (hot utility) and cooling (cold utility) which must be supplied to a given process. These minimum amounts are called the targets, and any process, no matter how well it is designed, must use at least the target amount of utilities. However, if a heat pump is introduced, energy use can be reduced below the target amount. The amount of reduction depends on the process, with some processes being more suitable for heat pumps than other processes.

Most of the heat pumps currently in use in industry are either MVRs (mechanical vapor recompressors) or TVRs (thermal vapor recompressors, also called steam-jet ejectors). This paper focuses on a third type of heat pump, the absorption heat pump (AHP). AHPs have existed for many years, with the well-known absorption chillers being one example of this. More recently, attention has begun to focus on the use of AHPs in industry.

In the late 1970s and early 1980s, 18 industrial AHPs with an output of 880 kW (3 MMBTU/hr) or greater were installed (RCG/Hagler, Bailly, 1990). Figure 1 shows an AHP installed in an ethanol distillation plant (Tokano and Yano, 1990). The AHP consists of four vertical falling film heat exchangers and several pumps, and has been operational since 1987. The heat of condensation of ethanol vapor at 79°C (174°F) is converted to 3.4 MW (12 MMBTU/hr) of heat at 112°C (233°F), which is used to reboil the column.

WHY AHPs?

MVRs require energy input in the form of shaft work from an electric motor, steam turbine, or engine. AHPs, on the other hand, operate using heat as the energy source. This heat source can be in the form of natural gas, steam, or process waste heat. AHPs can provide lifts up to 77°C (170°F) in a single-stage machine, and higher in a multi-stage machine. TVRs also use steam or heat as the input energy and are low in capital cost. Two major limitations of TVRs, however, are that their efficiency is low and they can only provide small lifts.

AHPs are not complex, consisting only of heat exchangers, pumps, and controls. Some electricity or shaft work is required to run the pumps, but the amount used is typically

Figure 1. AHP Applied to Ethanol Distillation

less than 2% of the heat pump load. For the AHP in Figure 1, for example, the pump energy input is 0.9% of the AHP energy output. Because pumps are the only moving part of an AHP, reliability can be expected to be high in the mature technology.

FINDING APPLICATIONS

Good AHP applications can be found using the same methods as for any other type of heat pump. Every process has a "pinch temperature" which is a characteristic of the specific process. In order to be appropriately placed, a heat pump must take in heat below the pinch temperature, and deliver heat above the pinch temperature (Linnhoff et al., 1982). An appropriately placed heat pump saves both hot utility and cold utility, whereas an inappropriately placed heat pump may actually increase utility requirements.

AHPs come in two basic types, the heat amplifier (HA), and the temperature amplifier (TA). An HA cycle requires a heat source above the pinch to drive it. A TA cycle requires a heat source below the pinch. Because most processes require cooling below the pinch anyway, a TA cycle can use the waste heat of the process. In this case the cost of energy to drive the heat pump is very low because besides the electricity needed to run the pumps, the only energy input to the AHP is waste heat. Figure 2 schematically shows the differences between the HA and TA cycles.

Figures 3 and 4 show the applicability diagrams for single-stage HA and TA cycles. The abscissa is the temperature of the waste heat which needs to be upgraded. The ordinate is the amount of lift that can be provided. There are two sets of curves on each graph, one for an AHP using lithium bromide (LiBr) as the working fluid, and one for an AHP using a newly developed aqueous alkali metal nitrate working fluid (Erickson, 1987). The nitrate fluid greatly extends the temperature range over which AHPs can operate. The assumptions used in constructing Figures 3 and 4 are:

- Minimum approach temperature of 5°C in each heat exchanger
- Three weight percent concentration differential between strong and weak absorbent
- Maximum absorbent temperature of 150°C for LiBr and 260°C for nitrate fluid
- Margin of 15°C to crystallization, both constant pressure and constant concentration

Details of applicability diagram construction can be found in Davidson and Erickson (1986). Note that the curves denote the maximum lift that can be achieved in a single stage, but any lower lift can also be served. Also, for higher lifts, staged heat pumps can be used.

CASE STUDIES

In order to assess the potential for AHPs in industry, it is necessary to estimate the capital cost of equipment. Therefore, a computer program was written to automate the AHP costing process. Given the temperatures of external fluid streams, the program first determines all the internal state points of the cycle. Then the log-mean temperature difference (LMTD) of each exchanger is determined, and the required exchanger area is determined based on a user-supplied heat transfer coefficient. The installed cost of the exchanger is then determined according to

$$\text{Cost (\$)} = 30{,}800 + 750 \times \text{Area}^{0.81}$$

for carbon steel (CS) shell-and-tube exchangers, where Area is in m^2, and

$$\text{Cost (\$)} = 30{,}800 + 1331 \times \text{Area}^{0.81}$$

for exchangers with CS shell and SS 304 tubes (Hall et al., 1990). For SS 304 plate heat exchangers, the uninstalled cost is given by

$$\text{Cost (\$)} = 635 \times \text{Area}^{0.778}$$

(Kumana, 1984), and the cost relationship for pumps is taken from Hall et al. (1988). An installed cost factor of 2.5 is applied to the plate exchanger and the pumps. A fixed cost of $300,000 is assmed for engineering, site work, etc. The unit can be sized either for a given waste heat input, a given heat output, or a given capital cost. Both the HA and TA cycle can be modeled.

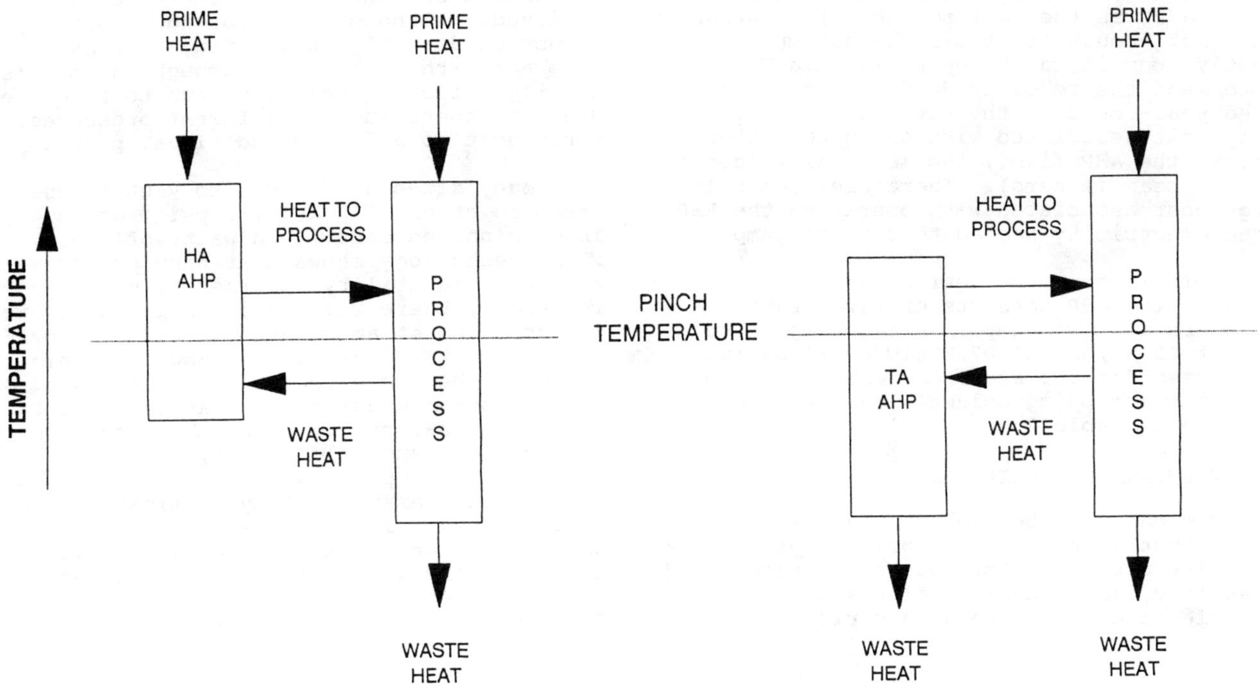

Figure 2. Comparison of HA and TA Cycles

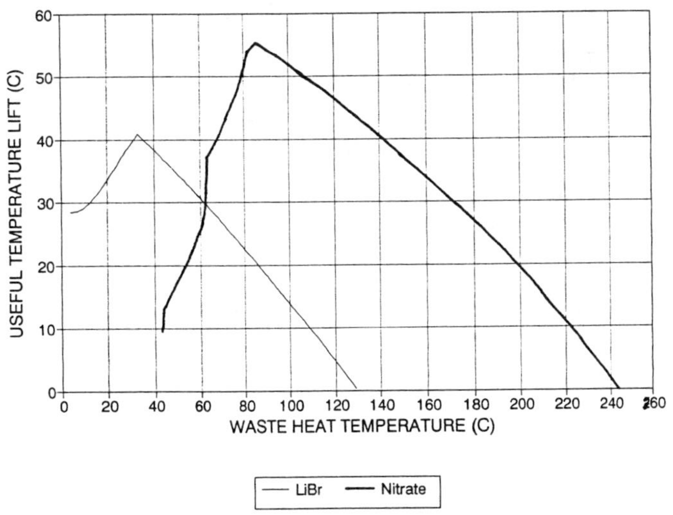

Figure 3. HA Applicability Diagram

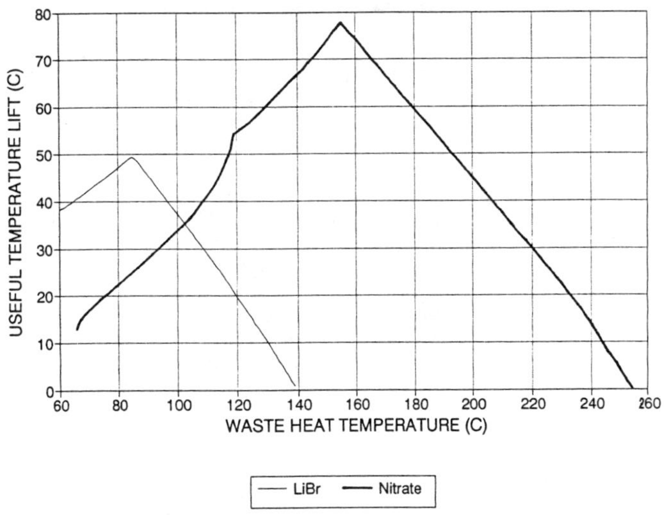

Figure 4. TA Applicability Diagram

In addition to performing the costing, the program also has the capability to optimize the AHP design. The augmented Lagrange multiplier method is used, and the objective function can be either the simple payback period, or the internal rate of return. The optimization is typically constrained by a maximum absorption fluid temperature, a maximum absorbent concentration, and crystallization limits. An optimization can be run on a microcomputer in about ten minutes.

Many different case studies have been analyzed, and many applications have been found with paybacks ranging from less than one year to three years. Following are examples of AHP applications in petroleum refining and steam compression.

BTX Units

It has been recognized that a heat pumping opportunity often exists in the benzene-toluene-xylene (BTX) plant of a refinery. One such opportunity was located at the solvent stripper of a U.S. refinery

(Tensa, 1989). An absorption heat pump can be run using the refinery's 375 kPa (40 psig) steam header as the heat source. This steam is not hot enough to reboil the column directly, but it can be upgraded in a TA mode AHP to meet the reboiler duty. This steam is vented year-round at the site, so there is no energy cost associated with using this steam to drive the AHP (i.e., the marginal value of 375 kPa steam is zero). Therefore, the only energy cost associated with operating the AHP is the electricity needed to run the pumps.

Table 1 shows the design parameters of the AHP. The AHP consists of five heat exchangers and two pumps. A representative value of \$2.91/GJ (\$3.07/MMBTU or \$4.02/1000 lb) is used for the steam currently used to reboil the stripping column. The AHP economics are shown in Table 2.

Absorption Steam Compressor

AHPs can also be employed in a site's utility system, rather than being applied to a particular process. For reasons of safety and operability, many chemical plants and refineries are organized as a series of individual plants. Unfortunately, this limits the amount of heat integration which can be achieved. Although the individual plants cannot be directly integrated, they can interact with each other through the site's utility system. Typically, two to five steam headers, operating at different pressures, carry heat to and from individual plants.

Many sites are forced to vent excess low-pressure steam (513 kPa (60 psig) or lower). This is not necessarily a waste of energy. Pinch technology shows that many processes require cold utility in order to operate. Therefore, there will be cases in which excess low-pressure steam cannot be recovered by further heat integration. However, this steam can be recovered using a heat pump. A heat pump can upgrade the heat to above the pinch temperature, reducing the requirements for both hot and cold utility.

Two different utility applications were investigated. In one case, it is assumed that waste steam at 547 kPa (65 psig) is available. A TA mode AHP is assumed, and units were sized and costed for various heat output temperatures. The heat output is held

TABLE 1. AHP DESIGN PARAMETERS AND COSTING

Exchanger parameters:

Exchanger	Shell Fluid	Shell Mat'l	Tubes Fluid	Tubes Mat'l	LMTD (C)
Generator	steam	CS	Alkitrate	SS	26
Absorber	Alkitrate	CS	bottom liq.	SS	37
Evaporator	water	CS	steam	CS	13
Condenser	steam	CS	water	CS	24
Recuperator	Alkitrate	SS plate	Alkitrate	SS plate	24

Summary of costs:

Heat exchangers:	349,438
Installation cost (1.5 times major equipment cost):	534,207
Fixed costs (engineering, site prep, etc.):	300,000
Installed cost	1,190,345

TABLE 2. ABSORPTION HEAT PUMP ECONOMICS

Heat source	: 10.4 kg/s 375 kPa steam (75.9 MMBTU/hr 40 psig steam @ 287°F)
Heat sink	: 10.1 MW @ 151°C (34.5 MMBTU/hr @ 304°F)
Savings	: 10.1 MW X \$.00291/MJ X 8000 hr/yr X 3600 s/hr = \$850,246/yr
Electricity cost	: 26 kW X \$0.036/kWh X 8000 hr/yr = \$7,488/yr
Net savings	: \$842,758/yr
Capital cost	: \$1,190,345
Payback	: 1.4 years

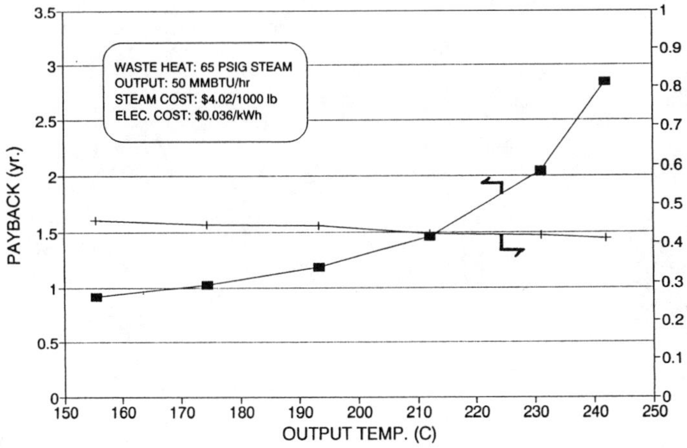

Figure 5. Payback period and COP as a function of output temperature for the TA mode.

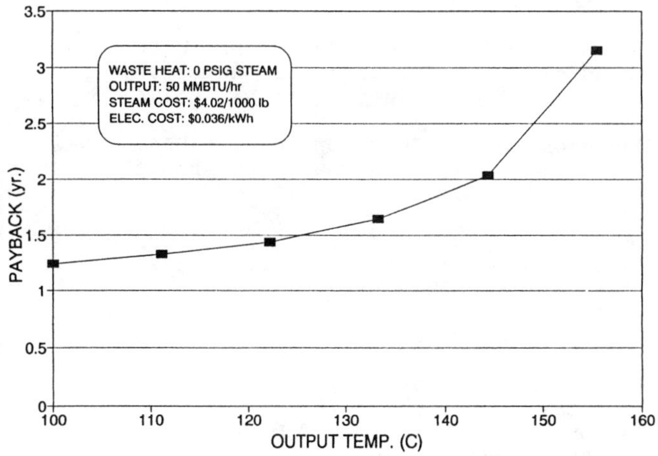

Figure 6. Payback period as a function of output temperature for the HA mode.

constant at 14.6 MW (50 MMBTU/hr). Figure 5 shows the payback period as a function of the heat output temperature. The heat output temperature is equal to the temperature of saturated steam produced by the AHP. The costs of steam and electricity are assumed to be $2.91/GJ ($3.07/MMBTU or $4.02/1000 lb) and $10.00/GJ ($0.036/kWh) respectively. The installed cost factor of 2.5 and fixed costs of $300,000 are included as before. The payback period increases as the heat pump lift increases. This is due to the fact that the temperature driving force in each exchanger is decreasing, thus increasing the size of the exchangers. The payback is under two years up to a heat pump lift of about 75°C (135°F).

It is interesting to note the contrast between this situation and that of a vapor compression heat pump (VCHP). The payback of a VCHP also increases with increasing lift, but primarily due to a decrease in COP. The COP of the AHP is virtually constant, as shown in Figure 5. The increase in payback period of the AHP is primarily due to an increase in capital cost.

In the other utility application, it is assumed that waste steam at 101 kPa (0 psig) is available. A direct gas-fired HA mode AHP is assumed. Figure 6 shows the payback as a function of the heat output temperature. In both of the above cases, the choice of waste heat temperature is arbitrary. Lower and higher waste heat temperatures are also possible. In both the HA and the TA modes, staged cycles can be used to get either higher lifts or higher efficiency when required.

DEVELOPMENT STATUS

Most existing industrial AHPs use lithium bromide (LiBr) as the working fluid. LiBr introduces several limitations to AHPs. The maximum temperature is limited to about 150°C (302°F) due to the corrosivity of LiBr (Holmberg and Berntsson, 1989). Copper-nickel or titanium is required for heat exchangers which contact the hot LiBr solution. Users have experienced varying levels of corrosion problems, with one user reporting a need to replace the titanium plates of a plate-and-frame exchanger every 6 months (Bouma, 1990).

The newly-developed aqueous nitrate working fluid, called Alkitrate, overcomes these limitations. Alkitrate extends the maximum working temperature to 260°C (500°F). Austenitic stainless steels, such as SS 304 and SS 316, have been used in pilot plant construction and performed satisfactorily. Therefore, Alkitrate will be useful in applications which already require stainless steel, such as food processing and corrosive services. Carbon steel has shown promise as a construction material in laboratory testing (DeVan and Wolf, 1988), but additional testing is desirable before carbon steel can be qualified for industrial use.

Alkitrate has been tested for a total of more than 500 hours of operation in a 10 kW pilot plant (Howe and Erickson, 1990), with a maximum recorded absorbent temperature of 277°C (530°F). An industrial scale demonstration project is scheduled to begin in 1991. The project is cost-shared between the Department of Energy and industry.

Although industrial AHPs may be considered experimental due to the fact that none have ever been installed in the U.S., it is also true that numerous companies in the U.S. and abroad have significant absorption technology experience. Given the relative simplicity of absorption machines, this technology should be able to enter the market quickly, once the first few demonstration projects have been completed.

ACKNOWLEDGEMENT

This research was partially supported by the Office of Industrial Technologies of the U.S. Department of Energy.

REFERENCES

Bouma, J.W.J., 1990. "Experience With a Heat Transformer in the Chemical Industry", IEA Heat Pump Centre Newsletter, **8**, no. 4, p. 13.

Davidson, W.F., Erickson, D.C., 1986. "New High Temperature Absorbent for Absorption Heat Pumps", ORNL/Sub/85-22013/1.

DeVan, J.H., Wolf, J.S., 1988. "Corrosion of Materials in Chemical Heat Pump Working Fluids", ORNL/TM-10896.

Erickson, D.C., U.S. Patent 4,652,279.

Hall, S.G., Ahmad, S., Smith, R., 1990. "Capital Cost Targets for Heat Exchanger Networks Comprising Mixed Materials of Construction, Pressure Ratings and Exchanger Types", Computers Chem. Engng., **14**, No. 3, pp. 319-335.

Hall, R.S., Vatavuk, W.M., Mathey, J., 1988. "Estimating Process Equipment Costs", Chemical Engineering, Nov. 21, p. 66.

Holmberg, P., Berntsson, T., 1989. "Some Aspects on Heat Transformer Optimization", IEA Heat Pump Centre Newsletter, **7**, no. 1, p. 9.

Howe, L.A., Erickson, D.C., 1990. "Proof-of-Concept Testing of Alkitrate™, Phase III Final Report", ECC-TR12, Energy Concepts Co., Annapolis, Maryland, USA.

Kumana, J.D., 1984. "Cost Update on Specialty Heat Exchangers", Chemical Engineering, June 25, p. 169.

Linnhoff, B., et al., 1982. User Guide on Process Integration for the Efficient Use of Energy, Pergamon Press Ltd., Oxford, U.K., pp. 88-116.

Priebe, S., Scheihing, P., 1991. Industrial Process Integration, Project Summary Report, presented at the Industrial Process Integration Conference, April 2-3, Houston, Texas.

RCG/Hagler, Bailly, 1990. "Opportunities for Industrial Chemical Heat Pumps in Process Industries", RCG/HBI 89-3077, prepared for Office of Industrial Technologies of U.S. DOE.

Tensa, 1989. Energy Conservation Study, Draft Report, EPRI project RP 2783-15.

Tokano, H., Yano, T., 1990. "Absorption Heat Transformer Saves Energy in an Alcohol Plant", IEA Heat Pump Centre Newsletter, **8**, no. 4, p. 4.

HEAT PUMP APPLICATIONS IN A CORN PROCESSING PLANT

Andrew S. McMullan
Linnhoff March, Incorporated
Leesburg, Virginia

ABSTRACT

The options for energy cost reduction in a corn processing plant have been identified in a systematic fashion using Pinch Technology. The opportunities include, heat recovery, process modifications, cogeneration and heat pumping. The analysis revealed several opportunities for heat pumping. This paper discusses the heat pumping opportunities identified and compares them to the competing alternatives.

INTRODUCTION

Heat pumping has long been recognized for its energy cost reduction potential. More recently, it has been recognized that heat pumping is only one of many possible energy cost reduction options, and that it is necessary to understand how a proposed heat pumping system competes with the other options before recommending installation.[1]

The use of Pinch Technology[2] as a thermodynamically sound approach to identifying heat pumping projects and comparing them to other options, such as heat recovery, process changes and cogeneration has been recommended.

This paper describes such an evaluation at a wet corn milling plant operated by American Fructose, Decatur Inc., Decatur, AL. The primary focus is on the types of opportunities which exist, the relative economics and other benefits available. Some details are also provided about a heat pumping system which will be installed. The paper is based on work completed under DOE's Office of Industrial Technologies Industrial Heat Pump Program. A full report is available.[3]

SITE AND PROCESS DESCRIPTION

The primary product manufactured at the facility is high fructose corn syrup (HFCS). The overall process consists of two distinct operations - milling of the corn to obtain a starch slurry in the mill and conversion of the starch to HFCS in the refinery. Figure 1 is an overall block diagram of the process.

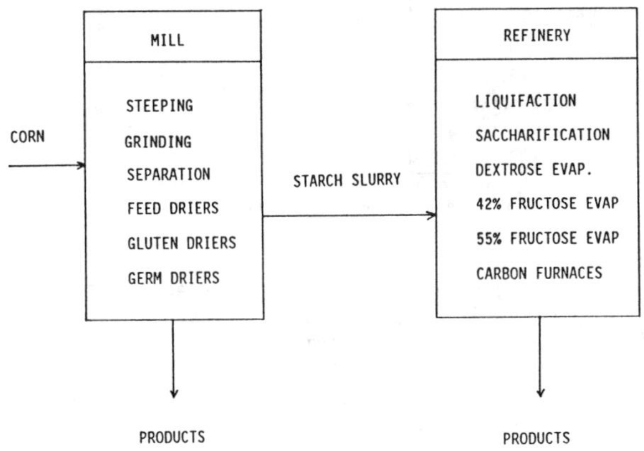

Figure 1 Block Diagram of HFCS Process

The case study presented will focus on the refinery processes. In the refinery, the starch slurry is treated thermally and enzymatically to produce fructose. The process is energy intensive due to the need for heating and cooling for carbon and ion-exchange treatment, and evaporation. Figure 2 provides more detail on the refinery processes.

The process uses 150 psig steam supplied by a coal fired boiler. Where necessary, pressure reducing stations are utilized. Also, a mechanical vapor recompression (MVR) evaporator provides part of the evaporation capacity.

The site has expanded over the years to the point where reductions in steam and cooling water use could have value in allowing future expansion without investment in utility equipment.

ANALYSIS

The general methodology followed[4] consisted of two stages; the Feasibility Stage and the Design Stage.

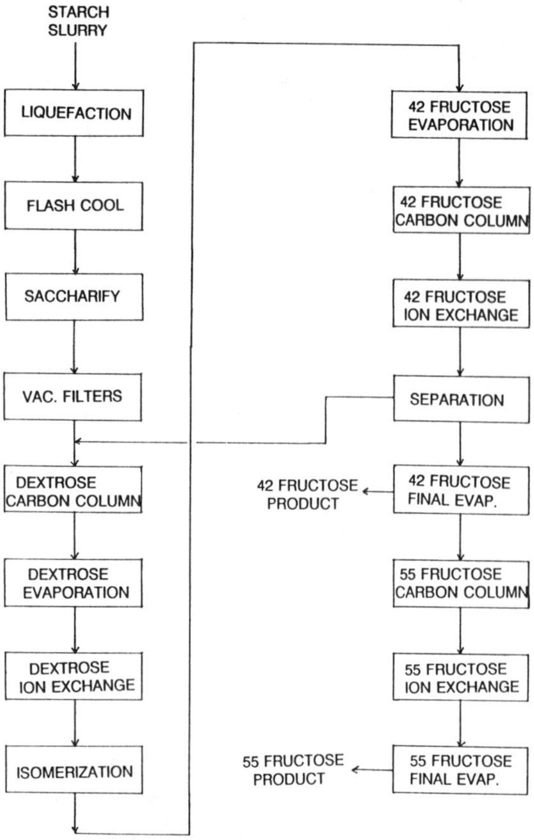

Figure 2 Block Diagram of Refinery Processes

The purpose of the feasibility stage is to identify potential projects and evaluate their relative technical and economic merit. The outcome of this stage is set projects which meet the process owners' acceptability criteria and which will be carried forward to detailed design and implementation.

The steps in the feasibility stage are shown in Figure 3. Typically, many possible projects will be identified, these will be evaluated and subjected to screening criteria. Those that remain will be evaluated in more detail, with the whole procedure being repeated until the most acceptable package is developed.

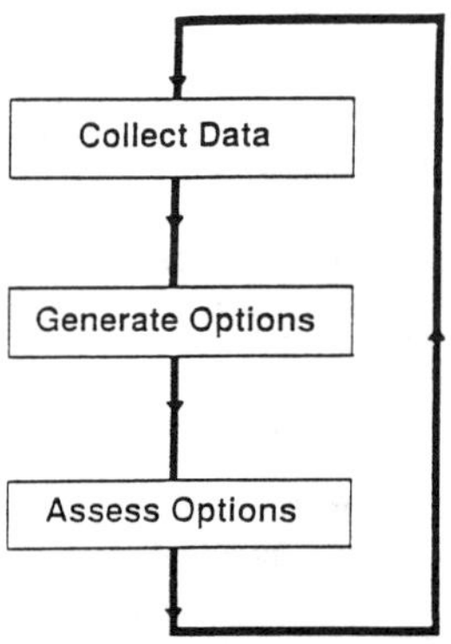

Figure 3 Steps in a Feasibility Analysis

A key to successful completion of the feasibility stage is a good means of generating options. Pinch Technology provides a systematic approach to this task. Its sound thermodynamic basis allows generation and evaluation of options to be completed in a rigorous and consistent fashion.

The design stage involves engineering the project proposals to the point at which installation can proceed.

For the process studied, use of pinch techniques to generate options revealed projects in the following categories:

- heat recovery
- heat pumping and process modification
- cogeneration

The project identified in each category are discussed below.

Heat Recovery

Identification of heat recovery projects has two stages; targeting and project identification. Targeting allows the scope for improved heat recovery to be determined before design of specific projects is undertaken. This is important because the magnitude of likely savings is determined before embarking on major design efforts. Also, the ultimate level of savings is defined so that the performance of the final project package can be judged against that target.

Table 1 compares the existing and target consumptions for the refinery processes. (The targets are calculated using the standard pinch concepts now well accepted by industry.) The scope to reduce energy consumption is 12%.

TABLE 1
ACTUAL VS TARGET ENERGY CONSUMPTION

Existing Consumption	645 units
Target Consumption	567 units
Scope to Save	12%

Note that the process already incorporates a heat pump, in the form of a mechanical vapor recompression (MVR) evaporator. The MVR supplies 66% of the total existing process heating duty, with the remaining 34% being supplied by steam from the boiler. As will be seen later, the 12% potential savings reduce the steam demand rather than the heat pump heat output. Thus, the process steam demand could be reduced by 35%.

Having determined that there is scope, the next step is to identify the specific projects that must be implemented to reduce process energy consumption to the target level. The reason for energy consumption in excess of the target is the existence of cross pinch heat transfer. Thus, to reduce steam consumption to the target level, it is necessary to identify projects which eliminate cross pinch heat transfer. This is accomplished by following a systematic procedure known as the Pinch Design Methodology.

This procedure was followed to identify refinery heat recovery projects. Typically, a long list of projects is generated initially. These are subjected to preliminary acceptability criterion, such as safety, operability and so on to develop a short list for further more detailed screening, following the general concept of the feasibility study described earlier.

Table 2 lists the finally agreed upon projects. This final project package achieves 76% of the initially targeted savings. The projects all save steam, reducing the boiler steam load by 29%. In addition, some projects also reduce the cooling duty, by a total of 27%. The overall project savings amount to $441,000/yr with an installed cost of $450,000/yr, resulting in a 1 year payback.

TABLE 2
REFINERY HEAT RECOVERY PROJECTS

1. Preheat Starch Slurry to Liquefication
2. Preheat Dextrose/Raffinate to Carbon Columns
3. Increased Preheat of Dextrose to MVR Evaporator
4. Increased Preheat of Fructose to MVR Evaporator
5. Preheat of Fructose to Final Evaporation
6. Preheat of Dextrose to Isomerization
7. Preheat of Syrup to Separators
8. Preheat of Water to Separators
9. Use of High Temperature Flue Gases
10. City Water Heating

It is worth noting that these types of heat recovery projects are common practice in HFCS refineries, there is nothing unusual about them. What is different about this collection of projects is the way in which the individual projects fit together in a package, obtaining a level of savings unlikely to be obtained by any other approach.

Process Modifications and Heat Pumping

The opportunities for heat pumping and process modification are guided by the plus/minus principle.[4, 5] This principle can be summarized as follows:

- a heat pump should take heat from below a pinch and deliver it above a pinch
- a beneficial process modification should result in more heat being available in hot streams above the pinch or less heat being required in cold streams above the pinch.

In this instance, the grand composite curve is the best tool to use for identifying opportunities because it clearly shows the quantity and temperature levels at which the process requires and rejects heat. Figure 4 shows the refinery grand composite curve.

The large 'nose' represents an excellent heat pumping opportunity because it represents a heat sink at a small temperature difference above a large heat source, with the pinch in between. In fact this 'nose' is actually already being exploited by the existing MVR evaporator as shown in Figure 5. However, this is not the only heat pump possibility. Each heat pump type (mechanical, chemical, thermal, etc.) has its own characteristic heat supply and rejection profiles. A knowledge of these characteristics, plus the grand composite curve, can lead to identification of other heat pumping opportunities.

In a water based process with significant evaporation requirements, there is often flexibility in selecting the temperatures and pressures at which the evaporation occurs, subject to product quality constraints. This flexibility can be exploited to improve overall process energy efficiency, and in a retrofit can result in certain equipment items to be unloaded, opening up the opportunity for increased production.

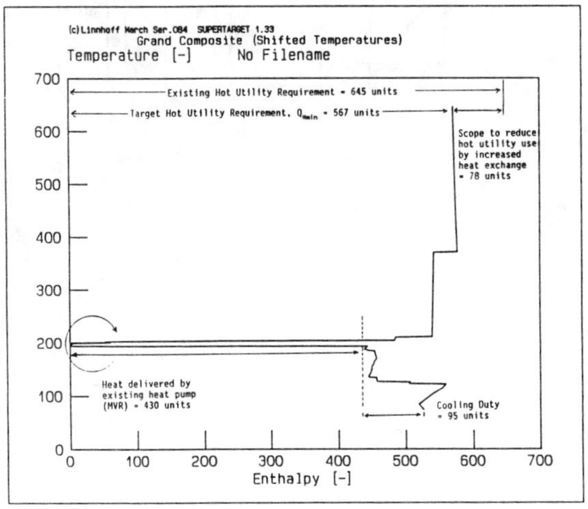

Figure 4 **Base Case Refinery Grand Composite Curve**

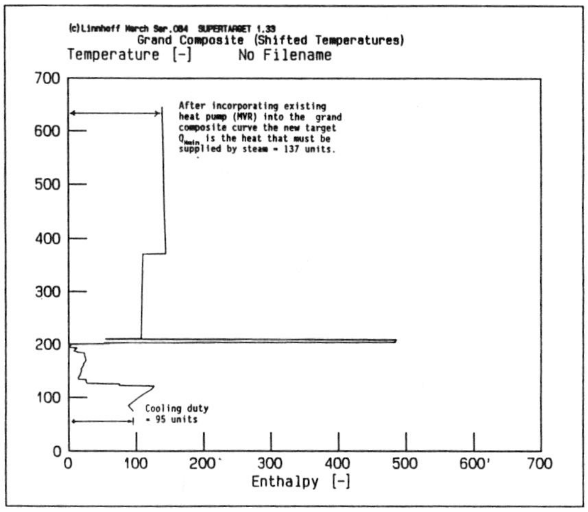

Figure 5 **Refinery Grand Composite Curve Incorporating Existing Heat Pump (MVC)**

Process Modifications. The following opportunities were identified:

1. Reducing the operating pressure of the multi-effect evaporator first effect(s) so that they operate entirely below the pinch (Figure 6).
2. Introducing a dextrose pre-evaporator which operates entirely above or entirely below the pinch (Figure 7).
3. Adding another effect to the multi-effect evaporators (Figure 8).

Table 3 summarizes the savings potential of each project. These savings are additional to those available from straight heat recovery.

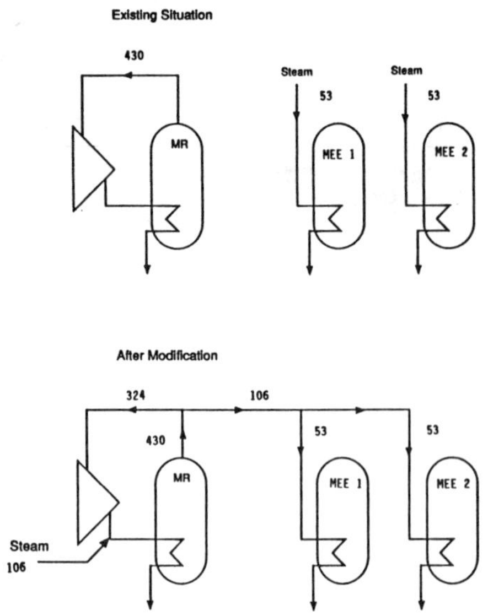

Figure 6 Impact of Reduced Evaporator First Effect Temperatures

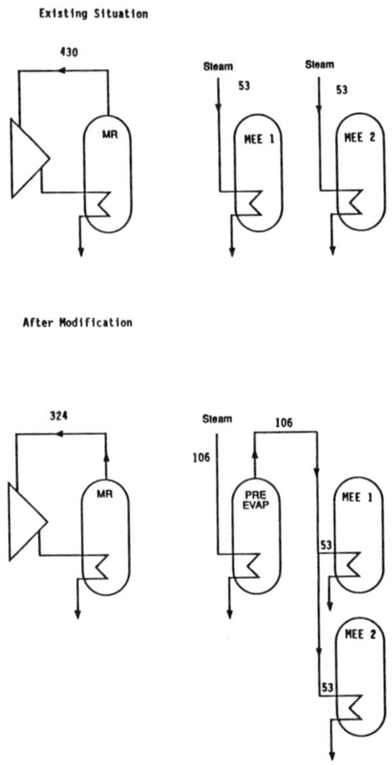

Figure 7 Integration of Pre-evaporator in the Refinery

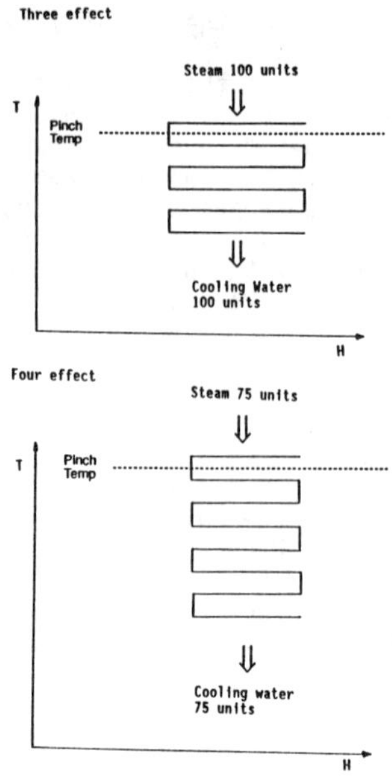

Figure 8 Impact of Additional Effects on a Multi-effect Evaporator

TABLE 3

SAVINGS AVAILABLE THROUGH PROCESS MODIFICATION

Project	Annual Savings $000/yr	Type of Utility Saved	Estimated Capital Cost $000	Payback yrs
1. Reduced First Effect Pressure	106	Power	--	--
2. Pre-evaporator	106	Power	850	8
3. Additional Effects	189	Steam	400	2.1

It is interesting to note that projects 1 and 2 would result in a reduction in the amount of heat supplied by the existing MVR compressor, relatively inexpensive heat. Project 3 results in steam savings which are of greater value.

Project 1 would have the effect of reducing the temperature driving forces for evaporation in each of the evaporator bodies, resulting in a need for additional exchanger area on each of the effects. This was considered unlikely to be economically attractive and was not considered further.

Project 2 would reduce evaporation in the existing dextrose evaporators, potentially allowing increased throughput.

Heat Pumping. Using the plus/minus principle, two primary heat pumping opportunities were identified:

1. Use of a thermocompressor to pump heat round the 'nose' on the grand composite (Figure 9).

2. Conversion of the existing multi-effect evaporators to operate with thermocompressors round the first effect.

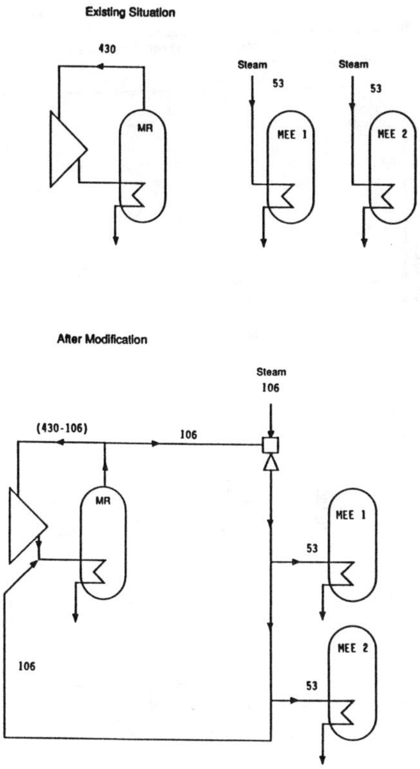

Figure 9 Use of Thermocompressor in Refinery

Project 2 adjusts the distribution of duties across each effect of the evaporator, as shown schematically in Figure 10. This results effectively in reduction of 'inappropriately placed' heat transfer across the pinch. Table 4 summarizes the savings potential of each project.

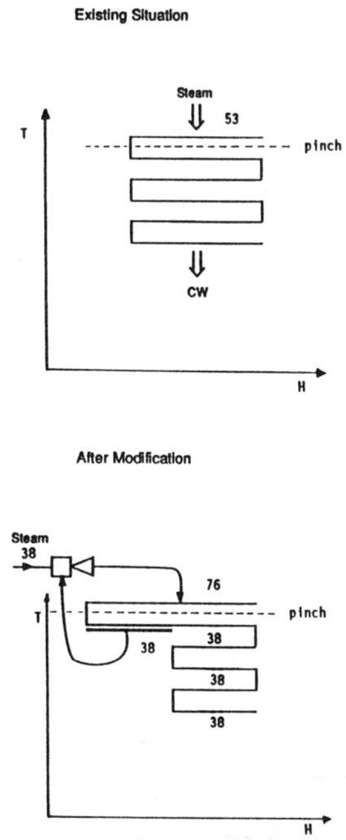

Figure 10 Effect of Shifting Loads and Thermocompression in the Evaporator

TABLE 4

HEAT PUMPING PROJECT SUMMARY

Project	Annual Savings	Type of Utility Saved	Capital Cost	Payback
1. Thermocompress round 'nose'	106,000	Electric Power	--	--
2. Thermocompress round first effects	189,000	Steam	510,000	2.7

Both projects result in reduced temperature driving forces in the first effects of the multi-effect evaporators. Therefore, the major capital investment is in additional evaporator surface area in the first effects.

An interesting consequence of Project 2 is that the heat loads in the remaining effects are reduced. This could allow the evaporator to 'run colder' throughout because of the over surface, or by sizing the first effect area correctly, evaporative capacity could be increased with only one new body. It may, therefore, be possible to gain extra capacity and energy efficiency in one project.

Combined Process Modification and Heat Pumping. In some circumstances, process modifications and heat pumping can result in the same benefit. However, it is also possible to combine opportunities to get a cumulative benefit.

In this instance, one particular combination offers the potential for energy savings and additional capacity in the evaporation system.

Such a scheme would involve introduction of a pre-evaporator plus thermocompression round the first effect of modified multi-effect evaporators. This is shown in Figure 11. The energy saving of $329,000 results through reduced power input to the existing MVR evaporator and reduced steam consumption in the multi-effect evaporators. The estimated capital cost is $1.75 MM resulting in a payback of about 5 years. While this is unlikely to be acceptable for an energy saving project, it does allow the refinery evaporation system to be expanded with only 3 new bodies. As four separate streams are being concentrated, this could be a very cost effective expansion route. Also, at increased production rates, the energy savings would be greater.

Cogeneration

The grand composite curve can be used to select cogeneration systems which best satisfy the process. Generally, it is best to include the entire site in this analysis because the utility system is common to all processing units. Clearly, selection of a cogeneration system is tied in to the design of the site evaporation systems because they account for a very large heating duty. In this instance, it is assumed that the site evaporation systems remain as they are now. If one of the process modification or heat pumping projects described earlier were to be installed, the cogeneration analysis would have to be repeated for the new utility balance.

Three basic system configurations were identified:

1. Modify boiler steam conditions to 600#, 750°F and install steam turbine with 150# extraction and 25# exhaust (Figure 12). (The boiler was designed to allow such modification.)
2. Gas turbine to supply heat to feed drying system (Figure 13).
3. Install small steam turbine to expand 150# steam to 15# to supply multi-effect evaporators (Figure 14).

Table 5 summarizes the cogeneration savings.

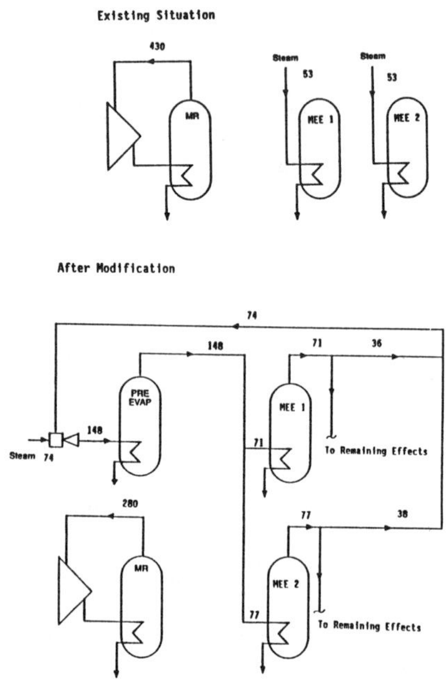

Figure 11 Equipment configuration with Shifted Loads, Pre-evaporation and Thermocompression

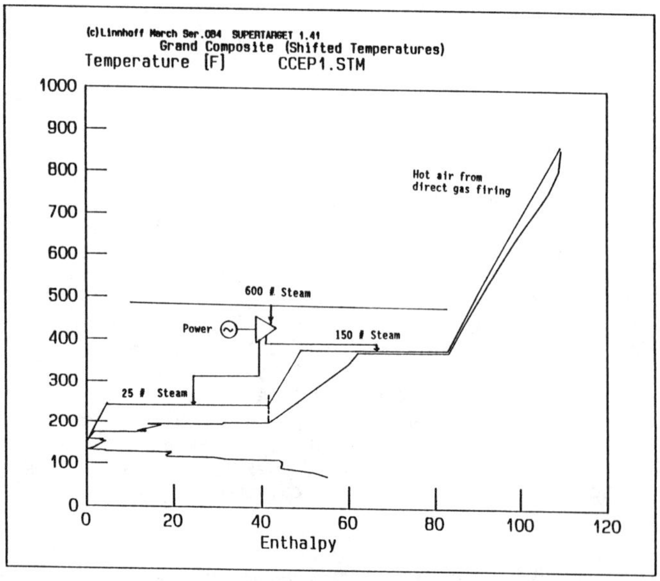

Figure 12 2-Steam Level Cogeneration System

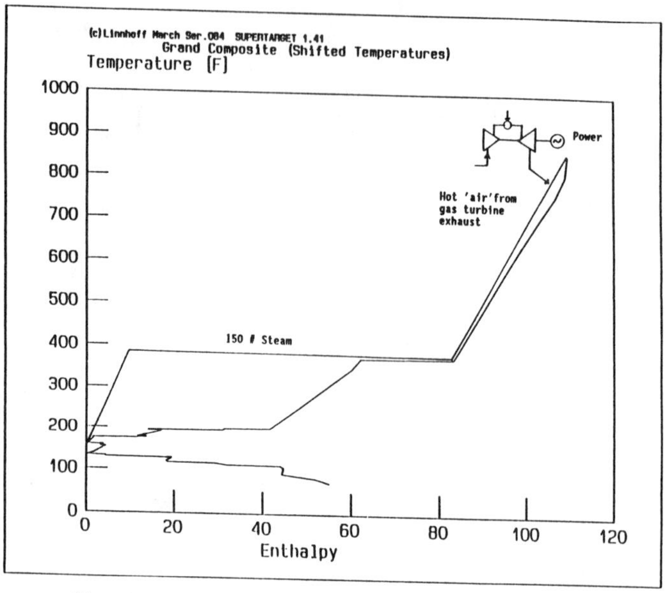

Figure 13 Gas Turbine Integration to Supply Hot Air for Drying

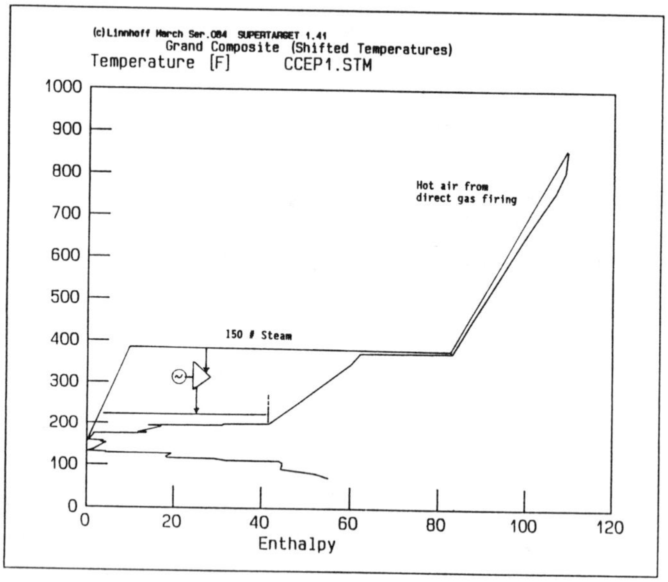

Figure 14 Single Level Steam Cogeneration System

TABLE 5

SUMMARY OF COGENERATION PROJECTS

Project	Saving $000/yr	Capital Cost $000	Payback years
1. HP steam, 150# and 25# distribution	550	4,000	7.3
2. Gas Turbine	800	3,500	4.4
3. Small Steam Turbine	157	345	2.2

The rather long paybacks of these projects could be expected, because the power to fuel price ratio at the site is 3.1:1.

Summary of Analysis

Table 6 compares the savings opportunities identified.

TABLE 6

SAVINGS BY PROJECT CATEGORY

Project	Saving $000/yr	Capital Cost $000	Payback Years
1. Heat Recovery	441	450	1
2. Process Modification	(1) 106 (2) 189		
3. Heat Pumping	(1) 106 (2) 189	510	2.7
4. Process Modification Plus Heat Pumping	329	1,750	5.3
5. Cogeneration	(1) 550 (2) 800 (3) 157	4,000 3,500 345	7.3 4.4 2.2

Implementation

At the present time, the plant is planning to install thermocompressors on their multi-effect evaporators (Heat Pump Option 2). This project offers good steam and cooling water savings at a reasonable payback. This can be considered as a first step towards implementing the total scheme. Integration of a pre-evaporator may be considered at a later date when debottlenecking of the evaporation system is required.

Implementation of the heat recovery projects is also being planned.

CONCLUSIONS

Energy cost reductions opportunities in a wet corn milling process have been identified using a structured approach incorporating Pinch Technology. Heat recovery, process modification, heat pumping and cogeneration projects are all applicable.

It was found that increased heat recovery was the most economic way to reduce energy costs. Steam consumption in the refinery processes could be reduced by 27% through implementing practical projects. The cost saving is $441,000/yr for a capital expenditure of $450,000.

The process examined was water based and involved a large amount of evaporation. Because of this, it was found that changes to process operating conditions (i.e., evaporation temperatures) could result in benefits similar to those available from heat pumping projects.

The use of heat pumps in wet corn mills is quite common, mechanical vapor compression and thermocompression being prevalent because of the low temperature lifts involved. It does, however, appear that opportunities exist to improve the overall performance of the systems already installed. These typically involve integration between systems which are usually considered stand-alone.

The economics of such projects can be acceptable (in the two to three year range). This payback is not as good as that for heat recovery projects.

Heat pumping and process modification projects can be combined to increase savings. In this instance, one such project would have the added benefit of allowing cost effective expansion of the refinery evaporation system with a minimum of new evaporator bodies. The payback of the project based on energy savings alone is around five years. If capital cost savings in an expansion were factored in, the payback would be much better.

Staff at the plant studied concluded that steam saving rather than expansion was their priority and are planning to implement a project to utilize thermocompressors on their multi-effect evaporators. This will save $189,000/yr for an investment of $510,000.

Energy savings are not the only benefit offered by heat pumps. Capacity gains and offset expenditure on steam and cooling water systems are benefits which should be factored into the evaluation.

Large cogeneration projects at the site are not economically attractive, due to the relatively low power/fuel cost ratio of 3.1:1. A simple induction turbine installation could save $157,000/yr at a payback of 2.2 years. However, the decision to utilize thermocompressors on the multi-effect evaporators has eliminated the sink for LP steam and this project is now not possible.

In general, evaluation of heat pumping options should be done in the context of the overall site heat and power balance, rather than the context of individual unit operations. This ensures that the heat pump system is designed correctly and does produce the expected benefits. Pinch techniques are valuable in assisting with such an evaluation.

ACKNOWLEDGEMENT

The case was based on work completed under DOE's Office of Industrial Technologies Industrial Heat Pump Program. A complete report is available from Office of Industrial Technologies, U.S. Department of Energy, 1000 Independence Ave., Washington, DC 20585.

REFERENCES

1. Gluckman, R. and A.S. McMullan, "Industrial Heat Pump Manual", EPRI, EM-6057, Oct. 1988.

2. Linnhoff, B., et. al., "User Guide on Process Integration for the Efficient Use of Energy", I.ChemE. (UK) 1985.

3. Linnhoff March, Inc., "New Industrial Heat Pump Application - American Fructose Inc.," Final Report DOE/ID/12790-2, April 1990.

4. Parker, S. and B. Linnhoff, "Heat Exchanger Networks with Process Modifications", I.ChemE. Annual Research Meeting, Bath (England), April 1986.

5. Smith, R. and P.S. Jones, "The Optimal Design of Integrated Evaporation Systems", Heat Recovery Systems and CHP, Vol. 10, No. 4, pp. 361-368, 1990, Pergannon Press plc.

MEASUREMENTS OF HEAT AND MASS TRANSFER COEFFICIENTS DURING ABSORPTION OF WATER VAPOR BY LITHIUM BROMIDE AND (Li, K, Na)NO_3 MIXTURES

A. Zaltash, M. R. Ally, and R. L. Linkous
Energy Division
L. N. Klatt
Analytical Chemistry Division
Oak Ridge National Laboratory
Oak Ridge, Tennessee

ABSTRACT

A knowledge of heat and mass transfer coefficients in heat pump fluids, plays an important role in the design of absorption machines. Heat and mass transfer coefficients as well as subcooling are measured for absorption of water vapor in (Li, K, Na)NO_3 and Lithium Bromide (LiBr) mixtures. The rate of absorption of water vapor is obtained from the difference in concentration of mixtures between inlet and outlet streams across the absorber. In situ concentrations of aqueous salt mixtures over temperature ranges between 353 to 408 K were calculated from density measurements. This technique of measurement is a convenient but not a very accurate (± 0.8 wt% salt) method of measuring the in situ salt concentration.

Results show that the subcooling at the absorber exit is not only a property of the fluid, but depends strongly on the process conditions. The subcooling in LiBr mixtures without additive is shown to vary between 2.2 and 24.3 K and the film heat transfer coefficient between 1365.2 W/m^2-K and 801.1 W/m^2-K respectively, depending upon process conditions. These empirical results will prove to be of value to heat pump manufacturers because they have a strong bearing on costs and performance. Heat and mass transfer coefficients in aqueous salt solutions are presented as a function of dimensionless numbers.

INTRODUCTION

In vertical falling film absorbers, the outside surface of a tube is covered by an absorbing liquid film and this fluid is cooled by flowing a coolant (cooling water) through the inside of the tube. The absorption of vapor by the liquid falling film involves simultaneous heat and mass transfer. In most absorption cycle equipment where falling film mass and heat transfer occurs, the Reynolds numbers of the liquid film are those generally associated with laminar flow. The cooling process influences the vapor pressure - concentration equilibrium between the two phases which produces a driving force for exchange of mass. The heat of absorption gives rise to temperature gradients which in turn affects the transfer of heat (Yih and Seagrave, 1980; Grossman, 1983).

The total heat transfer rate to the cooling water under steady-state conditions is,

$$Q = \dot{m}_w C_{pw} (T_{w2} - T_{w1}) = U_o A_o \Delta T_{lm} \qquad (1)$$

where U_o is the average overall heat transfer coefficient based on the log-mean temperature difference (ΔT_{lm}), $\dot{m}_w$ is the mass flow rate of cooling water, C_{pw} is the heat capacity of water, T_{w1} is the inlet water temperature, T_{w2} is the outlet water temperature, and A_o is the outside surface area of the tube.

The overall heat transfer resistance is the sum of the heat transfer resistance from the liquid film to the tube wall, the heat transfer resistance of the tube wall, and the heat transfer resistance from the tube wall to the cooling water represented as,

$$\frac{1}{U_o A_o} = \frac{1}{h_f A_o} + \frac{x_w}{k_w A_{lm}} + \frac{1}{h_i A_i} \qquad (2)$$

where x_w is the tube wall thickness, k_w is the thermal conductivity of tube material, A_{lm} is the tube log-mean area, and ΔT_{lm} is the log-mean temperature difference, defined as

$$\Delta T_{lm} = \frac{(T_h - T_{w2}) - (T_h - T_{w1})}{\ln\left[\frac{(T_h - T_{w2})}{(T_h - T_{w1})}\right]}$$

where T_h is the temperature of the vapor in the absorber. Equations (1) and (2) are based on the vapor temperature, T_h, rather than the interface temperature because the former is amenable to direct measurement. Additional reason for using T_h is that this temperature remains constant throughout the length of the absorber and we measure and report h_f as an averaged value over the absorber length not the local heat transfer coefficients. The difficulties in measuring the interface temperature (Yueksel and Schluender, 1988) which varies continuously along the absorber is easily avoided in our approach. The value of the heat transfer coefficient from the tube wall to the cooling water, h_i, for turbulent water flow is vital in determining the heat transfer coefficient from the film to the tube wall, h_f. The inside heat transfer coefficient, h_i, as a function of the flow conditions of the cooling water will be determined experimentally by the use of Wilson's Method of analysis (Bennett and Myers, 1974). These values will be compared to the values predicted by the Dittus-Boelter equation

$$Nu = 0.023\ Re^{0.8}\ Pr^{0.3} \tag{3}$$

The coefficient of heat transfer from the film to the tube wall, h_f, can be reduced to a dimensionless number called Nusselt number (Nu), defined as (Ramm, 1968)

$$Nu = \frac{h_f\ a_{or}}{k_f} \tag{4}$$

where k_f is the thermal conductivity of the film and a_{or} is the reduced film thickness given by (Ramm, 1968)

$$a_{or} = \left[\frac{\mu_f^2}{\rho_f^2\ g}\right]^{1/3} \tag{5}$$

The values of the film thickness (a_o) in laminar flow and the mean velocity of its flow (V_m) can be calculated from the following expressions (Ramm, 1968; Tailby and Portalski, 1960)

$$a_o = \left[\frac{3\ \Gamma_L\ \mu_f}{\rho_f^2\ g}\right]^{1/3} \tag{6}$$

$$V_m = \frac{\Gamma_L}{a_o\ \rho_f} \tag{7}$$

where ρ_f is the density of the film fluid, μ_f is the viscosity of the film fluid, and Γ_L is the film liquid mass flow rate ($\dot{m}$) per unit wetted perimeter (P) defined as:

$$\Gamma_L = \frac{\dot{m}}{P} \tag{8}$$

The flow conditions are characterized by the Reynolds number for liquids (Re_L) which is determined from V_m and the equivalent diameter, $D_{eq.} = 4\ a_o$, represented as

$$Re_L = \frac{4\ \Gamma_L}{\mu_f} \tag{9}$$

The empirical correlations for outside film heat transfer coefficients could be represented by

$$Nu = A\ Re_L^m\ Pr^n \tag{10}$$

where A, m, and n are determined experimentally. One such correlation given by Ramm (1968) for laminar flow of the film ($Re_L < 2300$) is

$$Nu = 0.67 \left[\frac{a_{or}}{L}\right]^{0.33} Re_L^{0.11}\ Pr^{0.33} \tag{11}$$

where L is the tube length and Pr is the Prandtl number for the liquid defined as

$$Pr = \frac{\mu_f\ C_p}{k_f} \tag{12}$$

The physical constants of the liquid (C_p, μ_f, k_f, and ρ_f) are taken at the mean temperature of the boundary layer defined as

$$T_f = \frac{t_f + t_w}{2}$$

where t_f is the mean temperature of the film and t_w is the mean temperature of the wall on the side where the film is flowing.

The absorption of vapor at steady state from the gas phase to the liquid phase can be described by two equations (Cussler, 1984):

$$N_1 = k_G\ (P_{10} - P_{1i}) \tag{13}$$

$$N_1 = k_L\ (C_{1i} - C_{10}) \tag{14}$$

where N_1 is the moles of solute 1 (water vapor) leaving the interface per unit time per unit area of interface, C_{10} is the bulk concentration of solute in the liquid phase, C_{1i} is the interfacial concentration of solute, P_{10} is the bulk pressure in the absorber, P_{1i} is the interfacial pressure, k_G is the individual gas-phase mass transfer coefficient, and k_L is the individual liquid-phase mass transfer coefficient. For interphase transfer it is easier to use an overall mass transfer coefficient defined by either of the following expressions:

$$N_1 = K_G\ (P_{10} - C_{10}\ H) \tag{15}$$

$$N_1 = K_L\ (\frac{P_{10}}{H} - C_{10}) \tag{16}$$

where H is a type of Henry's constant, K_L and K_G are the overall liquid-side and gas-side mass transfer coefficients given as

$$K_L = \frac{1}{\frac{1}{k_L} + \frac{1}{H\ k_G}} \tag{17}$$

$$K_G = \frac{1}{\frac{1}{k_G} + \frac{H}{k_L}} \tag{18}$$

Since the gas absorbed is considered pure (the vacuum pump is good down to 0.0133 Pa; therefore the absorbed vapor contains a trace amount of non-condensibles assumed to be negligible), the gas film resistance can be assumed to be absent. Under these conditions, the liquid film coefficient, k_L, is equal to the overall liquid-side mass transfer coefficient, K_L. The individual liquid-phase mass transfer coefficient can be reduced to a dimensionless number called the Sherwood number (Sh_L) defined as (Ramm, 1968):

$$Sh_L = \frac{k_L \, a_{or}}{D} \tag{19}$$

where D is the diffusivity. In practical calculations, equations of type (Ramm, 1968)

$$Sh_L = A \, Re_L^m \, Sc^n \, Ga^{-p/3} \tag{20}$$

are used, where the constants A, m, n, and p are determined experimentally, Sc is the Schmidt number, and Ga is the Galileo number which are given as:

$$Sc = \frac{\mu_f}{\rho_f \, D}$$

$$Ga = \frac{g \, \rho_f^2 \, L^3}{\mu_f^2}$$

The constants in Eq. (20) that are used by different authors are summarized in a table by Ramm (1968). From this table, the constants suggested by Krasheninnikov and Durasova (Ramm, 1968) are given as:

for the range $320 < Re_L < 1300$

$$A = 0.209 \, D_t^{0.27} \quad m = 0.05 \quad n = 0.5 \quad p = 0 \tag{21}$$

and for the range $1300 < Re_L < 2200$ are given as:

$$A = 0.0154 \, D_t^{0.27} \quad m = 0.42 \quad n = 0.5 \quad p = 0 \tag{22}$$

where D_t is the tube diameter, m.

EXPERIMENTAL APPARATUS AND TECHNIQUE

The schematic of Fig. 1 shows the apparatus used to obtain the heat and mass transfer data for the (Li, K, Na)NO_3 and LiBr mixtures. The main features of the apparatus are the generator, storage tank, absorber, and densitometers. The storage tank is used during charging of the solution in the generator. After charging the generator, distilled water is placed in the storage tank for use in shutdown or emergency dilutions. The generator is fitted with four 1200 W - 220 V, low flux heaters. Pressure measurements in this set-up are done by 0 to 3.447×10^6 Pa pressure transducers. The concentrated solution in the generator flows through the 18.85-376.99 rad/s (180-3600 rpm) gear pump with controller. The solution, when pumped, passes through the densitometer and the mass flow meter. This solution enters the top of the absorber.

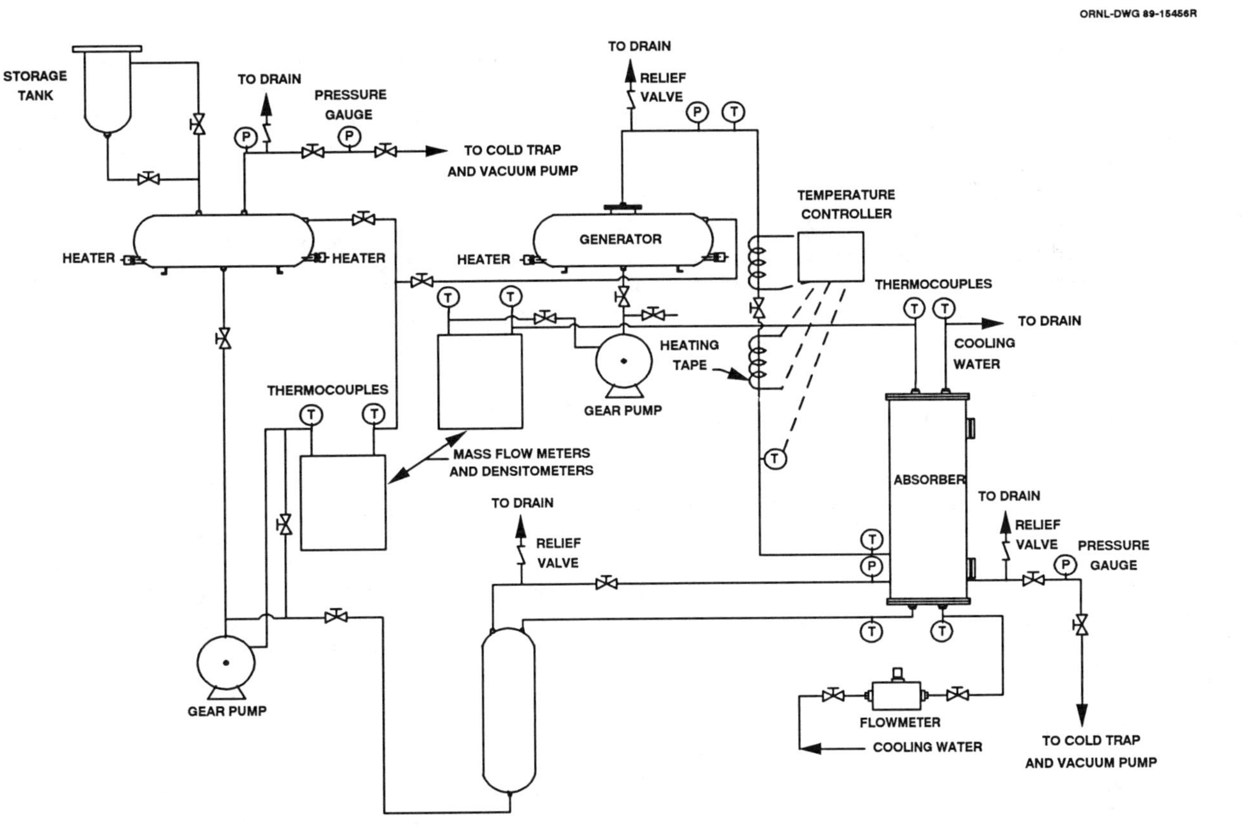

Fig. 1 Schematic of the apparatus used for heat and mass transfer experiments

The water vapor leaves the top of the generator and enters the bottom of the absorber. The vapor leaving the generator is maintained at a temperature higher than the fluid entering the absorber (approximately 2 K) to prevent condensation in the vapor line. At the absorber inlet, the concentrated fluid mixture is at equilibrium with respect to the water vapor in the liquid and vapor phases. As the fluid mixture flows down the outside of a vertical stainless steel tube (0.0127 m in diameter and 0.9144 m in length), it is cooled by cooling water flowing upward through the inside of this tube. This cooling reduces the equilibrium vapor pressure of water, giving rise to a driving force for the absorption of water vapor into the fluid. At the bottom of the absorber, the solute concentration is lower (more dilute) than at the top, and the difference is a measure of the amount of mass transfer of water from the vapor to the liquid phases. The rise in the temperature of the cooling water is a measure of the amount of heat removed during the absorption process.

The diluted solution flows through another pump and densitometer system similar to the one described above. This densitometer measures the concentration of the fluid exiting from the absorber. These data are subsequently used to calculate the average overall mass transfer coefficient and the individual and average overall heat transfer coefficients.

The system is purged daily, before starting experiments on each fluid sample. The system is checked for air tightness by overnight pressure and vacuum tests before and after experimental runs on each fluid. Another check of air tightness during the experiments is to reproduce the data collected several hours earlier.

DATA REDUCTION

The average overall heat transfer coefficient, U_o, is calculated from the total heat transferred to the cooling water, Eq. (1). In order to determine the outside heat transfer coefficient, h_f, one must calculate the other two heat transfer resistances (tube wall and tube wall to the cooling water resistances), Eq. (2). The tube wall resistance is calculated readily using the wall thickness of the tube (x_w), the tube log-mean area (A_{lm}), and the thermal conductivity of the tube material (k_w) which are known quantities. However, the inside heat transfer coefficient, h_i, for turbulent water flow must be determined either experimentally through the use of Wilson's method of analysis or by the use of Dittus-Boelter correlation, Eq. (3). In the Wilson's method of analysis, it is essential that all resistances other than the inside resistance be held constant during experimental runs at different cooling water velocities. This means that h_f which is a function of temperature drop over the fluid outside the tube (ΔT_o) must be held constant. In this study we used the Wilson's method to obtain the correlation for h_i and showed that the coefficients and exponents were almost identical to that in the Dittus-Boelter equation (Fig. 2). Hence we justify the use of the Dittus-Boelter equation in our experiments to obtain h_i values at various flow conditions.

The outside heat transfer coefficient could be estimated by Eq. (11). For LiBr mixtures all the physical constants needed to calculate Pr, Nu, and Re_L are known. However, for (Li, K, Na)NO_3 mixtures the thermal conductivity data over wide ranges of temperatures and concentrations are unavailable. In this paper, a constant value of 0.5362 W/m-K was assumed based on very limited data collected by Davidson and Erickson (1986).

The average overall liquid-side mass transfer coefficient, K_L, is calculated from the total amount of water absorbed by the liquid phase, Eq. (16). The liquid film coefficient, k_L, is equal to K_L based on the assumption that the gas film resistance is negligible. Equation (20) could be used to predict k_L. Of all the properties needed to calculate Sh_L and Sc, the diffusivity data of very concentrated mixtures (above 50 wt% salt in solution) are unavailable. However, the diffusivity could be roughly estimated by the use of Stokes-Einstein relation. Stokes-Einstein equation is a theoretical relation between diffusivity and viscosity, defined as (Bockris and Reddy, 1970; Ratcliff and Holdcroft, 1963)

$$D = \frac{k\,T}{6\,\pi\,r\,\mu} \tag{23}$$

where k is the Boltzmann constant, μ is the viscosity of the medium, T is the absolute temperature, and r is the molecular radius. It is more convenient to work with the ratio of diffusivities

$$\frac{D}{D_o} = \frac{r_o\,\mu_o}{r\,\mu} \tag{24}$$

where D_o and μ_o are the diffusivity and viscosity respectively, of water in a solution containing salts, LiBr or (Li, K, Na)NO_3, at infinite dilution. Equation (24) can be simplified further by the assumption that the ratio of ionic radii of salt mixtures at infinite dilution to salt mixtures at any concentration is equal to one, which gives

$$\frac{D}{D_o} = \frac{\mu_o}{\mu} \tag{25}$$

The diffusivity values obtained from Eq. (25) were used to roughly estimate the liquid phase mass transfer coefficient, k_L. The Nernst limiting values (salt concentration of zero) of diffusivity given by Robinson and Stokes (1965) at 298 K were corrected for temperature by

$$D_o(T) = D_o(298\text{ K}) \left[\frac{\mu_o(298\text{ K})}{\mu_o(T)}\right] \left[\frac{T}{298\text{ K}}\right]$$

and used in Eq. (25).

RESULTS AND DISCUSSION

The outside heat transfer coefficient, h_f, and liquid mass transfer coefficient, k_L, were obtained for LiBr and (Li, K, Na)NO_3 mixtures. These fluid mixtures flow laminarly down the outside surface of a 0.0127 m tube while absorption of water vapor occurs in the liquid film. Tables 1-4 show the flow conditions along with a comparison of experimental and predicted values for both heat and mass transfer coefficients. The average absolute deviation (AAD) used to compare these values is defined as

$$AAD = \frac{\sum |d_i|}{N}$$

where N is the number of data points and d_i is the relative deviation of experiments ($X_{i\,exp.}$) and predictions ($X_{i\,calc.}$):

$$d_i = \frac{X_{i\,exp.} - X_{i\,calc.}}{X_{i\,exp.}}$$

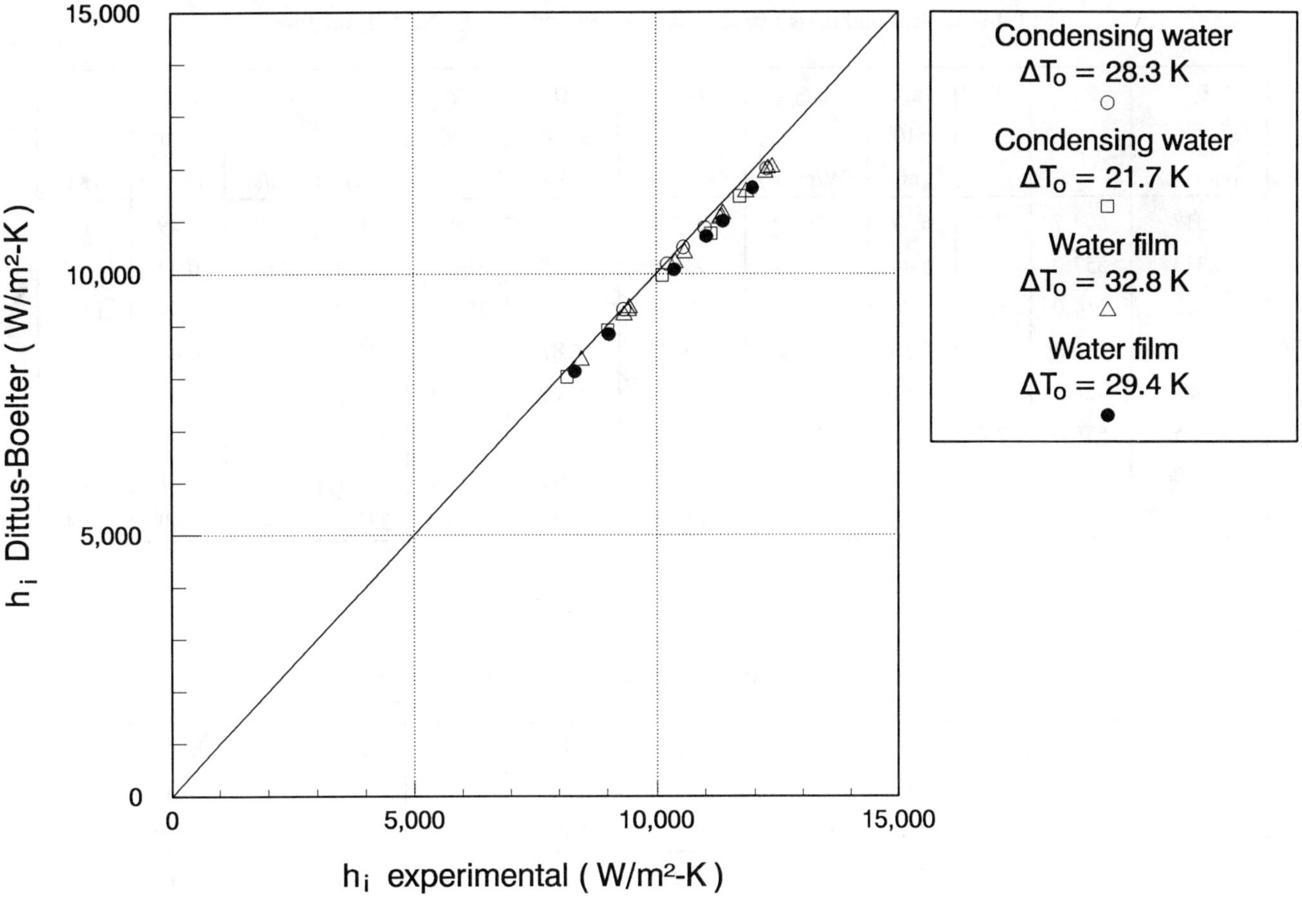

Fig. 2 Comparison of experimental and Dittus-Boelter values of heat transfer coefficient on the inside of 0.0127 m tube, ΔT_o is the temperature drop across the film

Table 1. Heat and mass transfer data for 59 wt% LiBr mixture

Γ_L (kg/m-s)	Re_L	Pr	a_{or} x 10^5 (m)	$h_{f\,exp.}$ (W/m^2-K)	$h_{f\,calc.}$ (W/m^2-K)	D x 10^{10} (m^2/s)	Sc	T_f (K)	$k_{L\,exp.}$ x 10^5 (m/s)	$k_{L\,calc.}$ x 10^5 (m/s)	$\Delta T_{sub.}$ (K)
0.313	684.9	7.8	5.11	1365.2	988.4	8.21	1392.9	362.8	2.18	0.53	2.2
0.307	581.0	9.1	5.62	1217.0	944.8	6.99	1887.1	356.0	1.22	0.48	7.9
0.306	494.5	10.7	6.18	1119.4	907.0	5.85	2600.2	348.6	0.76	0.42	14.1
0.301	426.3	12.4	6.73	983.2	872.2	5.01	3451.7	342.4	0.44	0.38	18.5
0.302	374.7	14.1	7.30	862.4	842.3	4.33	4505.8	336.9	0.24	0.34	20.8
0.297	336.9	15.6	7.74	801.1	820.4	3.90	5471.8	332.2	0.11	0.32	24.3

Table 2. Heat and mass transfer data for 59 wt% (Li, K, Na)NO_3 mixture

Γ_L (kg/m-s)	Re_L	Pr	a_{or} x 10^5 (m)	$h_{f\,exp.}$ (W/m^2-K)	$h_{f\,calc.}$ (W/m^2-K)	D x 10^{10} (m^2/s)	Sc	T_f (K)	$k_{L\,exp.}$ x 10^5 (m/s)	$k_{L\,calc.}$ x 10^5 (m/s)	$\Delta T_{sub.}$ (K)
0.317	604.7	7.2	6.17	1734.9	948.5	7.78	1950.5	350.6	3.75	0.49	11.3
0.313	557.0	7.7	6.44	1610.7	934.5	7.15	2265.9	345.9	3.03	0.47	12.2
0.312	514.0	8.3	6.74	1453.4	920.9	6.53	2655.6	341.0	2.69	0.44	13.2
0.310	466.3	9.1	7.13	1282.0	904.2	5.87	3214.5	335.4	2.30	0.41	13.9
0.308	428.3	9.8	7.51	1165.1	887.8	5.35	3803.9	330.7	2.03	0.38	14.9
0.315	578.6	7.4	6.28	1704.1	943.8	7.47	2087.0	349.1	3.22	0.48	12.1
0.229	473.3	6.6	5.88	1606.1	929.6	8.42	1675.0	350.1	2.85	0.51	12.7
0.311	462.8	9.2	7.18	1329.6	902.7	5.77	3301.5	333.8	2.46	0.40	13.4

Table 3. Heat and mass transfer data for 72 wt% (Li, K, Na)NO_3 mixture

Γ_L (kg/m-s)	Re_L	Pr	a_{or} x 10^5 (m)	$h_{f\,exp.}$ (W/m^2-K)	$h_{f\,calc.}$ (W/m^2-K)	D x 10^{10} (m^2/s)	Sc	T_f (K)	$k_{L\,exp.}$ x 10^5 (m/s)	$k_{L\,calc.}$ x 10^5 (m/s)	$\Delta T_{sub.}$ (K)
0.320	466.8	9.0	7.00	1708.1	912.7	6.38	2875.0	377.1	8.63	0.43	4.3
0.322	445.8	9.5	7.23	1672.5	903.4	6.00	3210.2	373.0	9.03	0.41	4.3
0.316	402.5	10.2	7.61	1529.5	886.2	5.48	3791.6	370.3	8.13	0.39	5.6
0.314	370.3	11.1	7.99	1404.9	872.3	5.01	4459.3	366.1	7.91	0.36	5.7
0.314	336.8	12.1	8.48	1301.8	854.7	4.51	5423.8	361.3	7.25	0.34	5.8
0.321	462.3	9.1	7.04	1682.7	911.1	6.31	2933.3	377.2	8.58	0.42	4.3

Table 4. Heat and mass transfer data for 87 wt% (Li, K, Na)NO_3 mixture

Γ_L (kg/m-s)	Re_L	Pr	a_{or} x 10^5 (m)	$h_{f\,exp.}$ (W/m^2-K)	$h_{f\,calc.}$ (W/m^2-K)	D x 10^{10} (m^2/s)	Sc	T_f (K)	$k_{L\,exp.}$ x 10^5 (m/s)	$k_{L\,calc.}$ x 10^5 (m/s)	$\Delta T_{sub.}$ (K)
0.307	184.3	19.6	11.7	1118.8	753.4	2.73	14579.6	391.4	3.29	0.23	14.1
0.306	151.3	23.8	13.3	949.8	724.0	2.17	22030.2	378.5	2.80	0.20	16.6
0.300	130.5	26.6	14.4	910.5	700.9	1.91	28245.6	377.5	1.10	0.18	21.9
0.298	118.4	29.0	15.2	839.6	686.0	1.74	33873.5	375.7	0.36	0.17	24.2
0.310	182.0	19.9	11.9	1157.8	751.6	2.68	15136.5	391.4	3.12	0.23	17.6
0.219	131.9	19.6	11.8	1123.5	725.1	2.71	14800.4	385.6	3.21	0.23	10.9

The film heat transfer coefficients, h_f, can be estimated from Eqs. (4) and (11) at various flow conditions. The overall average AAD for the LiBr and (Li, K, Na)NO_3 mixtures shown in Tables 1-4 are 14.2 % and 35.9 % respectively. The higher overall average AAD for (Li, K, Na)NO_3 mixtures could be due to the unavailability of accurate thermal conductivity data over the ranges of temperatures and concentrations studied. It should be noted that the tests on 72 wt% and 87 wt% (Li, K, Na)NO_3 mixtures were performed at higher average boundary layer temperatures, T_f, (Tables 1, 2, 3, and 4). The viscosities and the outside heat transfer coefficients of 72 wt% and 59 wt% (Li, K, Na)NO_3 mixtures were almost the same due to higher average boundary layer temperature of 72 wt% mixture.

The liquid film mass transfer coefficients, k_L, can be predicted from Eqs. (19) and (21) at various process conditions. The overall average AAD for the LiBr and (Li, K, Na)NO_3 mixtures shown in Tables 1-4 are 71.9 % and 87.4 % respectively. Two explanations for the large AAD values for mass transfer coefficients are: (1) the lack of accurate diffusivity data for moderately concentrated solution, hence, we had to estimate the diffusivities by Eq. (25); (2) the accuracy of measuring the in situ concentration was $\pm$ 0.8 wt% salt. In a typical absorption process the change in concentration between the absorber entrance and exit locations is about 3 to 4 wt%. Taking into account the $\pm$ 0.8 wt% error at each end, the total error in measuring concentration difference could be as large as 50 %; thus, the large AAD values are not totally unexpected.

The subcooling in LiBr and (Li, K, Na)NO_3 mixtures shown in Tables 1-4 signifies inadequate absorption which is defined as

$$\Delta T_{sub.} = T' - T_{out}$$

where T′ is the equilibrium temperature at the outlet concentration of the mixture and pressure of the absorber and T_{out} is the actual temperature of the mixture at the absorber outlet. The subcooling results in Tables 1-4 are shown to vary with the process conditions and are not just a property of the fluid. The higher vapor pressure driving force or lower liquid film flow rate (higher residence time) translates into more mass transfer, hence, lower subcooling (Fig. 3). It is interesting to see from Fig. 3 the steep slope for LiBr and the flatter curves for the three concentration of (Li, K, Na)NO_3 mixtures. The slopes can serve as a criteria of fluid performance. A fluid with steep slope is one that will lower the subcooling at a much faster rate per unit change of driving force, ΔP. Hence, Fig. 3 shows that LiBr is a better absorbent of water vapor than (Li, K, Na)NO_3 salt at higher temperature (higher ΔP), but may be poorer at low temperature (small ΔP).

CONCLUSIONS

Heat and mass transfer coefficients as well as subcooling were measured during the absorption of water vapor by (Li, K, Na)NO_3

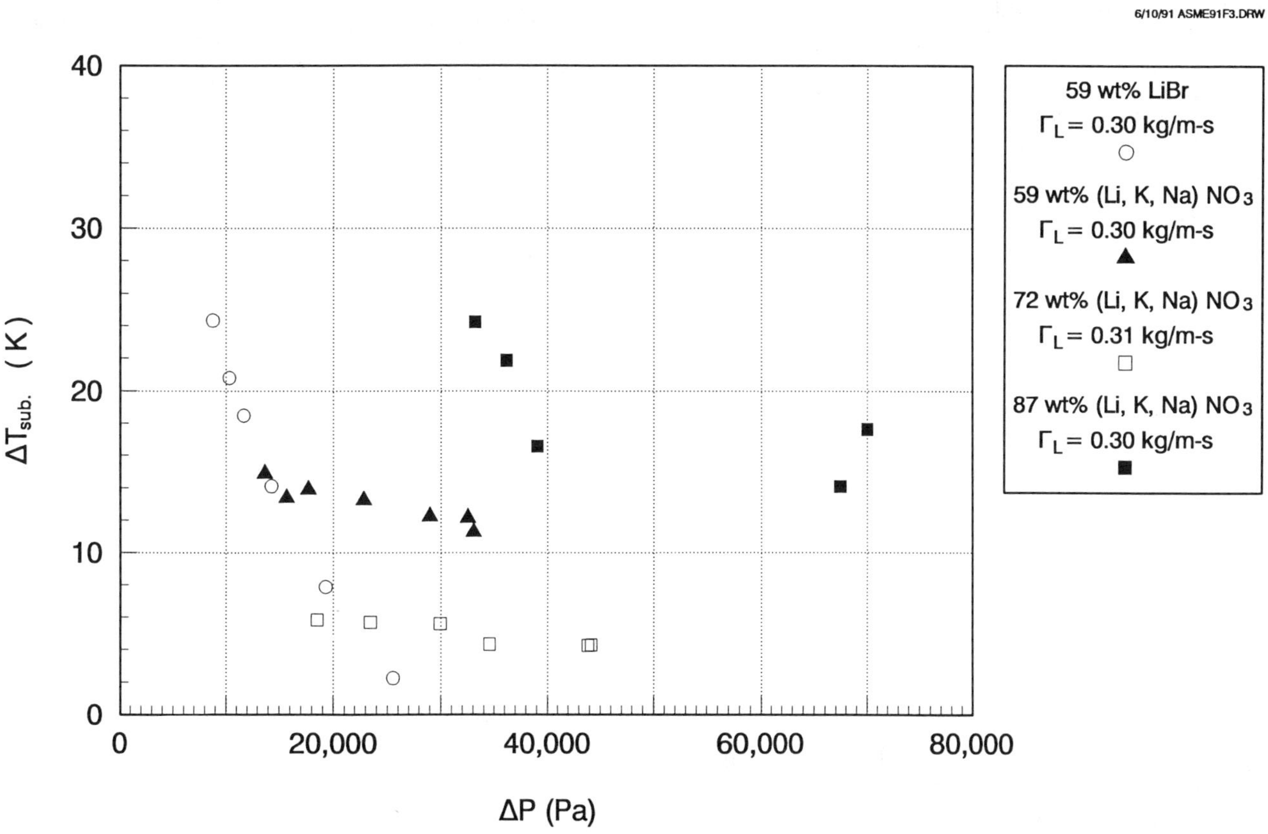

Fig. 3 The subcooling of LiBr and (Li, K, Na)NO_3 mixtures ($\Delta T_{sub.}$) as a function of vapor pressure driving force (ΔP)

and by LiBr mixtures in the absence of any additive. In predicting heat and mass transfer coefficients in aqueous salt mixtures, tentative dimensionless relations were presented. The outside heat transfer coefficient was defined through the Nusselt number, and the coefficient of mass transfer from the interface to the bulk of the liquid film was defined through the Sherwood number. The subcooling is a property of the fluid and depends on the process conditions such as liquid film flow rate (residence time or time of exposure) and vapor pressure driving force.

The in situ technique of measuring salt concentrations is convenient but not very accurate, being limited to $\pm$ 0.8 wt%. This is cited as the dominant reason why the measured mass transfer coefficients deviated significantly from those predicted by published correlations. However, for the type of salts examined and the high concentration and temperature of our experiments, the in situ method of concentration measurements were very convenient and gave consistent results. We hope this technique will be of value to future researchers.

ACKNOWLEDGMENTS

Research was sponsored by the Office of Industrial Technology, U.S. Department of Energy under contract DE-AC05-84OR21400 with Martin Marietta Energy Systems, Inc.

REFERENCES

Bennett, C.O. and Myers, J.E., *Momentum, Heat, and Mass Transfer*, 2nd ed., McGraw-Hill Book Company, New York, 1974, pp. 458-460.

Bockris, J.O'M. and Reddy, A.K.N., *Modern Electrochemistry*, Vol. 1, Plenum Press, New York, 1970, pp. 377-381.

Cussler, E.L., *DIFFUSION Mass Transfer in Fluid Systems*, Cambridge University Press, Cambridge, 1984, pp. 240-244.

Davidson, W.F. and Erickson, D.C., "New High Temperature Absorbent for Absorption Heat Pumps," ORNL/Sub/85-22013/1, July 1986, Oak Ridge National Laboratory, Oak Ridge, TN.

Grossman, G., "Simultaneous Heat and Mass Transfer in Film Absorption Under Laminar Flow," *International Journal of Heat and Mass Transfer*, Vol. 26, No. 3, 1983, pp. 357-371.

Ramm, V.M., *Absorption of Gases*, Israel Program for Scientific Translations, Jerusalem, 1968, pp. 278-306.

Ratcliff, G.A., Holdcroft, J.G., "Diffusivities of Gases in Aqueous Electrolyte Solutions," *Transactions of the Institution of Chemical Engineers*, Vol. 41, 1963, pp. 315-319.

Robinson, R.A. and Stokes, R.H., *Electrolyte Solutions*, 2nd ed. revised, Butterworths, London, 1965, pp. 513-515.

Tailby, S.R. and Portalski, S., "The Hydrodynamics of Liquid Films Flowing on a Vertical Surface," *Transactions of the Institution of Chemical Engineers*, Vol. 38, 1960, pp. 324-330.

Yih, S.M. and Seagrave, R.C., "Mass Transfer in Laminar Falling Liquid Films with Accompanying Heat Transfer and Interfacial Shear," *International Journal of Heat and Mass Transfer*, Vol. 23, 1980, pp. 749-758.

Yueksel, M.L. and Schluender, E.U., "Heat and Mass Transfer in Non-Isothermal Absorption of Gases in Falling Liquid Films," translated from German (ORNL/TR-90/4), *Waerme-und Stoffuebertragung*, Vol. 22, 1988, pp. 209-217.

AUTHOR INDEX

Ally, M. R. 83
Charoensupaya, Dhanes 19
Chen, F. C. 57
Desai, M. P. 39
Erickson, Donald C. 69
Gerlich, A. 49
Haubert, T. D. 11
Heenan, P. 57
Howe, Lawrence A. 69
Katipamula, Srinivas 1
Klatt, L. N. 83
Knocke, K. F. 49
Lambert, J. E. 11
Lavan, Zalman 19
Linkous, R. L. 83
Liu, Lianquan 19
Majumdar, P. 25
Mathiprakasam, B. 57
McMullan, Andrew S. 75
Mei, V. C. 57
Miles, Daniel J. 33
Mokadam, R. G. 39
Nguyen, D. C. 39
Nowakowski, G. A. 11
O'Neal, Dennis L. 1
Perkinson, Dennis 63
Sarwar, M. K. 25
Shelton, Sam V. 33
Shukla, Deepak 63
Sindermann, F. L. 39
Singh, Rajiv 63
Strauss, R. 49
Wigmore, D. B. 39
Zaltash, A. 83